Praise for POWERING CHANGE

"Tabitha Scott is a brilliant strategic thinker. When she tackles a challenge, she is miles ahead in finding solutions to bring balance to businesses in our ever-changing environment. As a former military officer, I often used the deep and insightful vision Tabitha refers to in her book and her techniques for accelerating adaptation, innovation, and productivity. She highlights the importance of action-oriented results with case studies showing how aligning your employees to the right roles, and allowing them to choose how to engage, helps retain those high performers. I believe everyone would benefit from reading this book and finding better ways to adapt, stay engaged, and work more effectively in their environment. A real eye-opener for executives who lead businesses."
—Ivan G. Bolden, Division Chief, U.S. Army

"*Powering Change* is an exploration into harnessing the inherent energy within individuals to revolutionize modern organizations. Grounded in science and bolstered by Scott's deep leadership experience, rigorous research, case studies, and personal reflections, this book offers a fresh perspective on fostering sustainable change and avoiding the ubiquitous burnout in today's fast-paced work environments.

Tabitha Scott provides leaders with an empowering guide to fostering positive cultures and unearthing unprecedented levels of engagement, productivity, and innovation. A compelling read for all who aspire to unlock their workforce's full potential, *Powering Change* is a visionary and pragmatic journey towards an attainable future, where businesses thrive on the renewable and limitless energy of human potential. Prepare to be inspired and equipped to power meaningful change in your organization."
—Gail Martino, Ph.D, NA, Lead Agile Transformation, Unilever

"Tabitha Scott unveils an original, modernized perspective that will change the way you lead. Having worked with her fifteen years ago when she was SVP of Innovation and Sustainability for Actus Lend Lease, I can honestly say the programs she created then have adapted with the times and are still delivering impactful results to this day. Her Human Dynamics Framework provides a fresh approach to solving the ever-present problems of burnout, motivation, and productivity. I recommend *Powering Change* for the leaders of today who are looking to be the leaders of the future."
—Patrick M. Appleby, President, WinnResidential

"*Powering Change* is an engaging book on the future of successful companies. Tabitha Scott has a sharp sense of how nature works, pivoting on the role played by change and energy in a flowing society that eliminates obstacles at every turn. Prominent in this natural design is the Growth curve (S-curve), its origin in the Constructal Law of evolution in nature, and how to use the Growth curve in the future. Tabitha delights us with a fascinating story of how we are connected to each other and with nature. *Powering Change* is rooted in wisdom and practical experience and is strongly recommended."
—Adrian Bejan, author of *The Physics of Life* and *Freedom and Evolution*, Distinguished Professor of Mechanical Engineering at Duke University

"The principles outlined in *Powering Change* have been purified in the crucible of Tabitha Scott's acute intelligence and her pioneering business and technology experience. Her wisdom, insights, and vision are laser-focused, practical, and aligned to the leading edge of business needs and survival. *Powering Change* not only represents the best of what I teach my graduate students about the innovation found in Earth's systems, but also the hope that we are already living among the answers to higher prosperity and healthier, lower-risk business growth."
—G. Dodd Galbreath, Graduate Director of Lipscomb University's Institute for Sustainable Practice and Bipartisan Public Servant to multiple Tennessee Governors and Mayors

POWERING CHANGE

THE CURRENT INSIGHTS SERIES

POWERING CHANGE

The Hidden Resource
to Unleash Potential,
Productivity, and Profits

Tabitha A. Scott
CEM, CDSM, CHTP

Copyright © 2024 by Tabitha A. Scott

All rights reserved. No portion of this book may be reproduced, stored in a retrieval system, or transmitted in any form or by any means—electronic, mechanical, photocopy, recording, scanning, or other—except for brief quotations in critical reviews or articles, without the prior written permission of the publisher. No representations, warranties, or guarantees are made with the publication of this book. The advice, strategies, and recommendations made herein may not fit your specific situation or circumstances. Neither the publisher nor author will be liable for any consequences of actions taken as a result of this book. While every attempt has been made to ensure scientific accuracy through a thorough review process, this book is not intended as a scientific resource. The reader should address scientific concerns with an academically credentialed physicist, scholar, or scientist.

Some names and identifying details have been changed to protect the privacy of individuals.

powering POTENTIAL Published by Powering Potential Media

Cover design: Ian Koviak and Alan Dino Hebel, Book Designers.com
Interior design and layout: Christy Day, Constellation Book Services
Interior graphic designers: Justin and Jessica Redmond, Samantha LaVoi

ISBN 978-1-7354940-4-3 (paperback)
ISBN 978-1-7354940-5-0 (eBook)
ISBN 978-1-7354940-6-7 (audiobook)

Library of Congress Control Number: 9781735494067

Printed in the United States of America

TABLE OF CONTENTS

Foreword by Heather Andre — ix
Introducing Your Competitive Edge — 1

PART I: SYSTEM OVERLOAD: THE CURRENT STATE — 5

Chapter 1: The Accelerating Pace of Change — 7
Chapter 2: Imbalance — 15
Chapter 3: The Burnout Crisis — 25

PART II: HUMAN DYNAMICS: POWER FROM THE SOURCE — 39

Chapter 4: Energy: The Catalyst for Change — 41
Chapter 5: The Growth Curve: Nature's Navigational Tool — 70
Chapter 6: Network Power: Welcome to the Wood Wide Web — 91
Chapter 7: Variation: Solutions for Resilience and Sustainability — 104

PART III: UNLEASHING POTENTIAL ENERGY — 121

Chapter 8: The AEM-Cube: Three Dimensions for Change and Growth — 123
Chapter 9: Executive Leadership Dynamics — 144
Chapter 10: Team Leadership Dynamics — 159
Chapter 11: Workplace Culture Dynamics — 173
Chapter 12: Purpose: The Fourth Dimension of Success — 189

Conclusion — 201
Be the Change — 201
Acknowledgements — 203
Sources Cited — 204

FOREWORD

by Heather Andre, SVP Customer Experience, First Horizon Bank

We've all sat through keynote speakers and change management "experts" who try to put a new spin on tired business concepts and advice, but who rarely teach us anything new about business and, more importantly, ourselves. This is *not* the case with Tabitha Scott. A two-day workshop she led for my department resulted in an epiphany about my way of working that I had long puzzled over but couldn't put language to. I immediately applied this realization to my daily life, with astounding personal and professional results. I knew then that Tabitha was onto something special.

As a senior vice president of marketing and customer experience who serves on the Diversity, Equity, and Inclusion Council at one of the largest banks in the southeastern United States, I'm no stranger to the instinct to try to outrun change. However, after having had the pleasure of working with Tabitha for several years, I have learned that the goal isn't to "manage" change, but to learn to flow with it by optimizing the resources we have already at our disposal.

How do we do this? It starts by realizing that our people are our competitive edge and our greatest resource. With Tabitha's Human Dynamics Framework, which was developed over many years of research and application, leaders learn how and why to assess, align, and adapt talent to what energizes them most.

One of the things I love about Tabitha's method is that it works for teams that are already vibing well in addition to those that are stuck or disconnected. With Human Dynamics, an unproductive team can be re-aligned to become productive, and a productive team can become even more so.

Powering Change points out what leaders forget, that employees' purpose coming to work each day is one of the most important metrics we should be tracking. Seeking to understand if our people feel seen, valued, and cared for takes a certain type of leader and culture. *Powering Change* helps the reader understand what it takes to create a culture that energizes people in a true, sustainable way.

I experienced Tabitha's infectious enthusiasm firsthand during a number of strategic engagements. From helping us devise and execute a program to revolutionize the customer experience to employing the tenets of cognitive diversity to design more productive team structures, Tabitha's approach discovers the hidden current already flowing within organizations and brings it to the surface.

People often mistakenly see the business and personal worlds as having completely different rules. Not Tabitha. She has an ability to humanize the business world in a way that creates the kind of culture that I have described here. A human-centered approach is the only way to genuinely understand your business, employees, and clients. Customers can only be as happy as your employees. Employees can only reach their highest potential when their work is properly aligned with their talents.

I've read many business books, and *Powering Change* gives a new and fresh perspective. Tabitha Scott is a business futurist because her truly innovative perspective helps leaders get unstuck and view problems (and their solutions) through a new lens. Her ability to challenge outdated management practices in a way that unlocks potential is truly rare, as is her way of tying the business world and high-performing teams to something we can all relate to; nature. By linking business cycles to natural cycles honed to perfection over billions of years, Tabitha helps people connect with concepts in a way that no other business book does.

In this world of unrelenting change, increasing burnout, and the pressure to constantly do more with less, *Powering Change* teaches us

FOREWORD

to tap into the unlimited resource we already have, the potential and power of our people. Our businesses and our world will be a better place for it.

INTRODUCING YOUR COMPETITIVE EDGE

Change management is a myth. There are too many unpredictable variables in the world for us to manage. Technological advancement has outpaced human adaptability, so just going faster is no longer enough. Employees are in a state of imbalance, and the "normal" ways of working and living from the past aren't coming back. Trust of employers is at an all-time low and burnout is at an all-time high. What worked for us in the past can no longer guide us into the future, and keeping up cannot be achieved with traditional business tactics.

What if I affirmed that you already have all the resources needed to unleash greater productivity, potential, and profits? What if the solution already flows within every organization and any leader can better their business by tapping into its hidden current?

It's true. This limitless, renewable, and free resource lies tucked within every person and has a track record dating back to prehistoric times.

Drawing upon decades of business experience, observation, and research, I developed the Human Dynamics Framework—a guide to help leaders navigate change and cultivate positive cultures where people thrive. This framework, grounded in physics and bioscience, harnesses the power and predictability of scientific laws and patterns that occur in the natural world to empower leaders to *flow with change*, to evolve the way they look at it, and recognize its foreseeable patterns throughout all living things, including businesses.

At the core of Human Dynamics is the realization that modern businesses are essentially living organisms, following the same principles of growth and interaction as natural systems. By leveraging billions of years of evolution, efficiency, and synergies, businesses can achieve greater success. Nature's networks, cycles, and variation have already proven resiliency in the face of constant, unpredictable change—and there is much to learn from them. The core principles that govern all living things can also be applied to every aspect of business and employee engagement.

When you read *Powering Change,* you will understand the core energy that drives people and how tapping into it provides a competitive advantage for modern businesses, converting burnout into engagement. When each person aligns with what energizes them most, the resulting synergy propels productivity, innovation, and profits. The key to unlocking success lies in leaders identifying those preferences, then aligning them with job responsibilities to unleash peak performance.

Stop chasing change. You'll never catch it.

Instead of chasing change, the key is to ride the predictable cycles of the current, harnessing the ever-present electricity that flows within your workforce. This current manifests as the energy and inspiration that drives ideas and accomplishments. It has the power to transform tired team members into connected, directed, and empowered contributors.

This book is the culmination of over 20 years of transformational executive, innovation, and energy leadership experience in various organizations. Rigorous, double-blind studies have also validated the positive results of the methods that follow. My intention is for *Powering Change* to inspire you to unlock your own power and forge meaningful connections with those around you. By sparking a positive movement within ourselves, we can ignite change through better leadership in teams, organizations, and communities. If you're ready to unleash your company's potential, productivity, and profits, this book is for you.

Part I
SYSTEM OVERLOAD: THE CURRENT STATE

Chapter 1

THE ACCELERATING PACE OF CHANGE

*Change is inevitable.
Except from a vending machine.*
—ROBERT C. GALLAGHER

The world of business has always been like riding a roller coaster full of twists and turns and ups and downs. But today, the pace of change is accelerating so fast, it's more like we're racing down the information superhighway. The rules and protective restraints of the past can no longer keep us safely on course.

Leadership in the Fast Lane

In 2013, I had the rare opportunity to sit trackside next to a pit crew at the Phoenix International Raceway during the NASCAR Camping World Series race. Completely unfamiliar with the sport, we were there to show support for Erik Jones, whose sponsor generously offered to raise awareness for our nonprofit, Wake Up Narcolepsy, by placing its logo on his vehicle during the race. When Erik made history that night by becoming the youngest winner of a NASCAR event at age 17, our website reached more people than all our prior campaigns combined.

Sitting in the pit was fascinating! I could hear the audio streaming live to Dave Jones, Erik's father and coach, who sat beside me. There were spotters perched throughout the grandstand, high above the race, who could see the big picture of the entire track in real time. They called in opportunities and warnings in constant communication back and forth. Erik listened intently and reacted immediately to avoid disaster and capitalize on openings to get ahead.

In the workplace, our "spotters" are the internal and external strategists who are constantly anticipating what's next in the "future of work." They see the big picture and provide navigational context. Which companies are leading the pack and why? Are there regulatory, equipment, environmental, or safety threats to consider? Where are opportunities emerging? In addition to these moment-by-moment predictions of the future, companies must also react to ongoing feedback from markets, consumers, and regulatory authorities.

At the raceway, the roar of the car stopping at the pit to our left rattled our ribs and amped up our adrenaline. I felt my heart race while tires were tossed over the wall and replaced in mere seconds. My son and I were transfixed by the pit crew's perfect synchronization as they simultaneously anticipated and analyzed the services needed on the vehicle and adapted instantly to the needs at hand. There was no time to deliberate. Instead, they conducted repairs, added fuel, and completed balancing in the exact moment needed for maximum efficiency.

I think of the many "pit crew members" it takes to run a business effectively: Information Technology (IT), Operations, and Human Resources (HR) come to mind as first responders. As quickly as processes are established, they are already under review for improvement or replacement to adapt. Scott Galloway, professor at NYU and business thought leader, affirms, "Take any trend—social, business, or personal—and fast-forward 10 years. Even if your company isn't living in the year 2030 yet, the pandemic has spurred changes in consumer behavior and markets." Change is a constant for any business, especially with the ongoing need to upgrade technology, collect data, personalize communication, and decentralize decision-making.

The "10-Year Effect" rings true today across everything from delivery services, online shopping, remote work opportunities, and new technology adoption. I personally experienced this when writing the IT strategic plan for a Fortune 500 retailer just prior to and during Covid-19. Like most retailers, we accelerated mobile application capabilities, curbside pickup, and delivery services and shifted to working remotely. While a long-term vision was still needed to guide major shifts in software and hardware, the technical changes generated by the urgency of the pandemic created too much uncertainty to make fixed plans too far into the future.

How Did Modern Leaders Get Here?

It takes a lot of energy to operate at this pace! How did we end up in this racecar and what can we do about it?

Think of leadership in the past as a comfortable commute—the destination was clear and there was time to react to unforeseen obstacles as they arose. Plans were predictable, budgets were fixed, and the primary purpose of meetings was to identify variances and then redirect actions back to the predetermined destination. It was a linear way of leading, with cause-and-effect planning and predictable outcomes—and in its day, it worked.

In 1943, Thomas Watson, president of IBM, predicted, "I think there is a world market for maybe five computers." At that time, the influence from the Industrial Age was still eminent. Innovation in the early to mid-1900s was centered on making systems more centralized, efficient, and productive. Companies identified what worked, then replicated it as effectively as possible to scale and mass-produce. Our utilities providers, mainframe computers, factory operations, financial accessibility, and myriad other systems sacrificed customization for efficiency.

An excerpt from Henry Ford's *My Life and Work* notes, "The salesmen…were spurred by the great sales to think that even greater sales might be had if only we had more models. It is strange how, just as soon as an article becomes successful, somebody starts to think that it would be more successful if only it were different. There is a tendency to keep monkeying with styles and to spoil a good thing by changing it."

Ford represented a type of leadership that saw change as a distraction that would dilute efficiency in production. That type of thinking may have prospered when there was little competition and predictable market conditions, but over time, growth in technology began to offer more customization and distribution options.

Gordon E. Moore, a co-founder of Intel, first described the pace of technological change in 1965. According to "Moore's Law," the transistor count on integrated circuits would double roughly every two years. It has held largely true for nearly 60 years now, even though he only predicted it to hold up for about 10 years. According to Our World in Data, the number of transistors that fit into a microprocessor reached over 10 billion in 2017 (the number was under 10,000 in 1971).

Soon after Moore's prediction, the Information Age was born in the mid-1970s when game-changers Steve Jobs and Bill Gates founded Apple and Microsoft. They had different ideas about business than their predecessors, Watson and Ford. They felt that the greatest innovations would come from the power of the people rather than huge organizations, so they put computing ingenuity into the palms of everyone's hands. It was the beginning of a "bottom up" revolution where customers took control of the wheel and began driving product innovation.

What early computers and smart devices lacked in capacity, they quickly made up for in accessibility and creativity while processing and storage capabilities caught up. As of 2022, the International Telecommunications Union estimates 2.7 billion computers are operational in the world (not counting an additional 6.6 billion[1] smartphones). By 2025, an estimated 38.6 billion smart devices will be collecting, analyzing, and sharing data.[2] What's more, today's hand-held smartphones have more computing power than was available to NASA when they landed on the moon![3] The explosion of digitization is revolutionizing the ways we live, work, and connect with each other.

> The leisurely commute of the past has been replaced by racecars speeding at 200 mph, and linear math no longer adds up to predictable outcomes. There are simply too many changing variables.

When you're moving at that pace, just one little social media post, one unforeseen lawsuit, or one virus can send a company into a tailspin and out of the race. Gone are the days when a great crisis-response public relations team could predictably protect a company's image.

With the chaos of unbridled, organic growth and the structure of specialization through business processes, the structures that were originally developed to help companies get organized are now holding them back. According to Dee Hock, founder of Visa, an organization that can succeed in today's environment is a "self-organizing, self-governing, adaptive, non-linear, complex organism, organization, community or system, whether physical, biological or social, the behavior of which harmoniously blends characteristics of both chaos and order."

The Constants: Chaos & Complexity

Thriving within an environment of accelerating change and unpredictable market, social, and technological disruptions requires a tectonic shift in thinking akin to chaos theory, first introduced in the context of business management by Tom Peters in the 1980s. To help leaders prepare for rapid environmental and technological changes, chaos theory focuses on the unpredictability in occurrences and behaviors. It regards organizations as complex, dynamic, non-linear, and co-creative systems. Their future performance cannot be predicted by past and present events. In a state of chaos, organizations behave in ways that are simultaneously unpredictable (chaotic) and patterned (orderly).

Even in nature, systems gravitate toward complexity. As kids, my sister and I each took responsibility for a few rows of the family's garden to plant and tend the seeds of our choice. I wasn't as dutiful as she was about keeping the weeds at bay. Random plants and flowers sprouted incessantly in my rows. The longer I waited to hoe them out, the more chaotic my little garden became and the more effort it took to regain stability. Sometimes, the fruits I planted were completely overtaken by the unplanned dandelions and grasses that seemed to emerge from nowhere.

Organizations are the same way. The more they grow, the more highly specialized their business lines become and they must constantly pluck out unpredictability. While these centralized "silos" allow for efficiency and scalability, they lack the fluidity and adaptability to handle the erratic and random nature of today's business environment. Peters' chaos theory reinforced Jobs' and Gates' visionary view of the customer: The only way to adapt quickly in an ever-changing environment is to seek out and embrace the roots of change itself.

But it's hard to embrace something that outpaces our mind's own evolutionary ability. Digital marketing experts estimate that we see up to 10,000 ads a day, which translates into 3.6 million new inputs per person each year and creates systems overload on a conscious and subconscious level. Our minds have not evolved as quickly as technology, and it's changing the ways we interact. In fact, a study from Microsoft[4] states that our attention span has dropped by a third since the year 2000. While a goldfish can hold its attention for nine seconds, humans can only hold theirs for eight seconds! Because today's pace of change has *already surpassed* our ability to keep up, finding ways to drive faster is no longer enough.

With chaos comes complexity. Simply put, a complex system is made of many parts that interact with each other in multiple, unpredictable, and nonlinear ways. Examples include communication systems, social structures, software, electrical systems, social media, transportation, competition, corporate cultures, and many others. With so much complexity, predictable modeling and planning are moot. Today's companies are highly variable, complex systems that require equally complex leadership, willing to respond and adapt fluidly to unpredictability as it comes.

Break the Rearview

In recent years, even before the pandemic, the accelerating pace of change left many feeling off-balance and burned out. Now, in its aftermath, "systems overload" doesn't even begin to describe the fatigue people are feeling as they struggle to adapt amidst so many unpredictable twists and turns. Senior Managing Director Stephen Kulinski of Nashville's high-growth CBRE office notes, "We employ progressive leaders here, and many still ask when things will return to 'normal' (pre-Covid). We need to stop looking in the rearview mirror—we can never go back. We need to find a way to adapt quickly to this new world and keep moving forward."

But how can leaders find their edge without creating even more disruption? Start by breaking the rearview mirror. Around 20 years ago, automakers began augmenting rearview mirrors with digital cameras. According to General Motors, the change provided a clear view behind the vehicle, with no obstructions of passengers, headrests, or the vehicle's roof and rear pillars. Additionally, it improved the field of vision by 300%, or roughly four times greater than a standard rearview mirror.

NASCAR also began replacing rearview mirrors in its next-gen racecars in 2021 to reveal a more accurate perspective with digital cameras instead. Chris Buescher, one of the first drivers to experience the real-time rearview cameras in the NASCAR Cup Series test at Daytona International Speedway, commented, "The rearview camera is something that is really neat…you can actually see quite a bit more than you're used to."

What worked for leaders in the past cannot guide them into the future, and adaptability at today's pace cannot be achieved with traditional business methods. While the ride is accelerating, leaders can find their edge by leveraging the scientific laws of physics and principles of ethology, physics, chaos theory, and complexity as the new age models for management. By harnessing the power and predictability of scientific laws and patterns that occur in the natural world, modern business leaders can tap into the energetic turbine that transforms stagnant businesses into well-oiled, self-propelling machines. Buckle up, everyone—the ride is in motion.

Chapter 2

IMBALANCE

*I'm so busy, I don't know whether
I found a rope or lost my horse.*

—MARK TWAIN

Remember playing on the teeter-totter (or seesaw) as a kid with your friends? It was a long, thick plank attached horizontally to a pole that ran underneath. At rest, gravity would pull the plank down to one side or the other, but when a child sat on each end simultaneously, they could play a game by scooting closer or further away, based on weight, to achieve balance. The joy of the ride was pushing up from the ground and leveraging the weight of your friend to fly effortlessly upward. Then, the other person returned the favor, sending you back to the ground again. The cycle would ebb and flow, up and down, depending on the actions of your playmate. Sometimes it was a gentle, smooth ride and other times it was a wild, unexpected series of jolts.

Our interactions, both internal and external, are like the teeter-totter. They provide forces that impact people around us in positive or negative ways, creating a constant balancing act. When we bring someone down, they expect an apology to regain homeostasis. While playground equipment provides an obvious, physical representation of "ups and downs," the personal imbalance people experience in the modern world is created through a variety of forces. In the context of work, it is often felt, but may not be physically visible until being off-balance has occurred for so long that it manifests into burnout.

One must always be mindful of which variables are increasing their state of balance or what may be throwing them off-balance (causing mistrust, anxiety, fear, or any number of negative emotions). To protect ourselves and our companies from imbalance, we must carefully consider the factors within our work and home cultures that provide negative impacts to our balance and make frequent "scoots" in one direction or the other to avoid an unexpected crash.

Like managing change itself, performing a balancing act is nothing new. The norms and paradigms from the past were developed to cope with variances in day-to-day occurrences. What's different today is that the validation of truth, boundaries of work, and confidence in expectations are no longer clear. It makes us anxious, stressed out, and uncertain about our current state and our future.

Root Causes of Imbalance

Maintaining balance in the workplace looks much different today than even a decade ago. Events no one could have predicted have forever changed the priorities of leaders and employees. Upon researching the catalyst, evidence points to common elements that can suppress the general workforce.

Disconnection

Disconnection can be due to a variety of factors: lack of communication, physical separation, lack of trust, lack of purpose, unclear direction, and more. It also comes from dysfunctional operating structures such as placing people in roles that don't suit their natural preferences. Studies published in *Harvard Business Review*[5] confirm that aligning employee strengths with their work increases profitability 14% to 29%. From an energetic perspective, this practice allows workers to focus on areas that naturally resonate with what motivates them individually, meaning they produce at a higher output with less effort. Misalignment requires individuals to exert greater energy to align with something outside their natural preferences, creating the feelings of being off-balance and fatigue. Over time, ongoing disconnection between a worker and their role leads to burnout.

Lack of Purpose

There's no debate that Covid forever changed the way we perceive our jobs and ourselves, and "success" is no longer defined by dollars alone. In 2022, a record 50.5 million people in the U.S. quit their jobs—the highest quit rate in history. Burnout was cited as the main reason for leaving, at 40%, according to the Bureau of Labor Statistics.

After reconsidering their quality of life and what's really important, people are unwilling to return to unfulfilling jobs. The younger generations, in particular, expect purpose-driven and highly flexible work, as evidenced through the findings of the Deloitte Global 2022 Gen Z and Millennial Survey. Mental health and well-being are high on their priority list, as is a sense of ownership and connectivity.

According to Gartner in 2022, 65% of workers felt that the pandemic made them rethink the importance that work should have in their lives. Further, 52% began to seriously question the purpose of their day-to-day jobs. Purpose provides the connection between vision and execution; without it, work is something to be tolerated rather than enjoyed.

Values & Internal Bias

Workers engage more powerfully when workplace values align closely with their personal values. Corporate and team culture play a major role in demonstrating values in action. Sometimes misalignment is easy to recognize, such as with someone who is always on time for meetings when the overall culture is to drop in late. Other times, the lack of alignment isn't obvious until pointed out, like when Outlook sends that weekly report from your mobile device that shows you've been spending more and more time working on the weekends.

Additional areas for friction between what we believe and what we must do include industry regulations, government requirements, or internal biases such as religious expectations, education, socio-economic bias, and individual life experiences. Collectively, those beliefs develop into a personal list of "shoulds." Many of us are highly driven perfectionists. Can you relate to this self-talk? *I should be changing my approach based on potential stockholder reactions. I should be making adjustments based on worker needs. I should be more present at my child's school events. I should be more fit…should eat healthier…*the list goes on and on.

Over time, if our own values and purpose, as informed by our internal biases and external forces, are not aligned with those "shoulds," and if we don't recognize and protect ourselves, we end up feeling drained or off-balance. I playfully refer to this self-destructive behavior as "shoulding" on ourselves.

I spent much of my adult life internalizing this type of imbalance, locking it neatly in boxes to prevent distractions or slowing down to process it. Like many executives, I put pressure on myself to be the perfect leader, fundraiser, athlete, friend, and mom. It took half a lifetime to realize it was impossible to be all things to all people. Unfortunately, I didn't recognize how severe my imbalance was until its effect on my health became eminent.

Blurred Work-Life Boundaries

The Cambridge Dictionary defines a traditional view of work-life balance as "the amount of time you spend doing your job compared with the amount of time you spend with your family and doing things you enjoy." But it's not that simple anymore. The borders between "work," "home," and "doing things you enjoy" have been blurred in recent years. Gartner now defines work-life balance as "an aspect of employee well-being related to their ability to manage both personal and professional responsibilities with adequate time for rest and leisure." Further, Gartner found that U.S. employees value work-life balance even more than health benefits.[6]

The blur between work and life began in earnest with the integrated capabilities of mobile devices that brought real-time calls and messages of all types into the palms of our hands. That 24/7 accessibility forever changed the clearly delineated, traditional separation between work and home life. Covid was a critical tipping point for work-life *imbalance*. As we literally took our offices home with us, separation became impossible. No doubt, the world was in survival mode during lockdown, but the aftereffects of the global work-life imbalance linger on.

Hybrid Work

In 2021, Owl labs found that **55% of respondents work more hours remotely than at the physical office**, making it even more difficult to have personal boundaries or unplug. While many of us have worked remotely for decades, according to the U.S. Census, the Covid pandemic drove more than a third of Americans out of their workplaces to work from home. Whether we're at the office, the kitchen table, or some combination thereof, our surroundings impact our state of balance.

In March 2022, Robert Half, a global recruiting firm, released a survey that revealed 50% of U.S. workers would rather resign than be forced back to the office full-time. On the other hand, some workers relished a return to the workplace. **People are energized differently, and the key is for leaders to understand which scenario uniquely energizes each employee and leverage that preference.** A blanket approach is no longer viable. We cannot undo the experiences of working from home and the impact it has had on work styles and preferences. We must recognize that remote, on-site, or hybrid work environments are significant contributors to the quality of each worker's life and personal balance.

Quiet Quitting & Firing

The prevalent and overall imbalance in the workplace can be understood through the lens of the summer 2022 trend, #quietquitting. Triggered on TikTok by Zaiad Khan, a 20-something engineer, who defined the term as "You are still performing your duties, but you are no longer subscribing to the hustle culture mentally that work has to be our life. Your worth as a person is not defined by your labour."

While Zaiad was not the first to articulate this sentiment, his video came during a social tipping point, especially amongst Millennials and Gen Z-ers who place values and work-life balance at a premium. Instead of living to work, they work to make a living. They reject the notion that they should go "above and beyond" for their employers if the effort is not reciprocated. New personal boundaries are going up, with no sign of retreat. And while this may sound individually empowering on the surface, being emotionally detached from work contributes to personal and corporate imbalance.

On the flip side, managers are leveraging the lack of desire to return to the office and expend extra energy as an opportunity to follow suit with "quiet firing." In quiet firing, managers create a hostile working environment so that people will quit on their own, allowing the company to avoid paperwork or severance pay, and indirectly thin the herd.

This method is nothing new. It is simply being acknowledged and observed more publicly because the pandemic made it easier for employees to go away quietly without inter-office disruption. Ayalla Ruvio, associate professor of marketing at Michigan State University, has researched the trend and said, "The phenomenon has long existed on an individual basis. Recently, though, some companies have adopted it as a corporate 'strategy' to reduce headcount, circumvent layoff costs, and avoid broadcasting unpleasant financial news."

The Truth & Trust Vacuum

The key, root causes of corporate imbalance are further inflamed by the complete breakdown of both truth and trust in the workplace. One such casualty of this troubling reality is Kevin, who worked in communications at a Fortune 500 corporation for eight years. He was an advocate for developing young leaders and brought insights for human design to the company to improve both internal and external experiences. Each year, he had outstanding performance reviews. As a result, Kevin was right on track to land the VP of Communications role when his boss retired at the end of the year. They had discussed the transition a number of times.

The promotion was so important to him that Kevin delayed starting a family so he would be available to put in the 60 to 70 hours a week the position required. He didn't mind terribly because he was proud of the positive changes he was contributing to the workplace. His job was a large part of his identity; it fueled his creativity, and it gave him a sense of purpose.

Without warning, Kevin was one of 47 other mid-level managers cut from the company. Each person was read the same script by an HR rep. "We have had to make some difficult changes. Unfortunately, your role has been eliminated. By the end of this meeting, you will no longer have access to our VPN or files as they are company property. This termination is nothing personal. You didn't do anything wrong. We just needed to reduce costs…" Kevin barely heard a word further over the ring of shock that filled his ears at a shrill pitch, rendering whatever else followed a jumble of incomprehensible static.

In exchange for signing off on legal documents and a non-compete agreement, he was reminded how fortunate he was to receive a couple months of severance pay (some folks didn't get any). The irony was not lost on Kevin that as communications manager, he had worked hard to spin the image of his employer as a family-friendly, progressive, and all-around great place to work, despite what he often saw happening behind the curtain. A quarter of his life left a trail on his office Outlook of contacts, family birthdays, and volunteer activities that were now the property and perusal of someone else.

How could they not have warned me? How could they show such little regard for someone who has given everything to contribute to this company's success? Why did I make myself so vulnerable? Why didn't I create a backup plan?

Kevin's willingness to go "the extra mile" for any company ever again had been destroyed. It was a wake-up call, he thought: Start a family, take care of my own needs, and treat work as a job, not the top priority in my life. Not anymore.

Kevin is not a Gen Z-er, but he is one of millions who decided to prioritize life over work. Though his decision was made in the aftermath of a gut punch, he is part of a growing group of people who are mistrustful of corporate culture, and with good reason. Lingering uncertainties from the pandemic make trust in the workplace more important than ever. Things are not going back to the way they were pre-Covid, and new management practices have forever changed.

To perform at a high level of engagement, workers need to feel authentically trusted by their managers and employers. A perfect example of this is the debate around working from home. Insisting that workers be in the office full-time can be interpreted as a lack of trust. While leaders have good intentions to foster collaboration and control, forcing employees to go against their preferences for managing work-life balance can cause resentment and dissatisfaction.

Truth is another critical value in jeopardy within corporate cultures. Americans broadly agree that knowing what is true and what isn't is an increasingly tough task. Without a common source for information, how are companies supposed to navigate communications? Navigating the "truth crisis," the latest challenge of the Information Age, is an imperative for businesses and organizations of all kinds, which are already experiencing more uncertainty and even less trust—including within businesses themselves.

There's a difference between objective truth and subjective perspectives. Sometimes it's difficult to separate these, since so much in the world is subjective and ambiguous and falls into a gray area—particularly when it comes to human relationships. This also means resisting the temptation to bend the story. For employees and internal audiences, this has to do with clear and upfront company policies and working expectations.

It also is important to conduct transparent surveys, going beyond impersonal, quantitative annual assessments to embracing qualitative conversations. Without understanding the "why," organizations cannot effectively address the "what." For external audiences, this means ready access to sustainability policies, hiring plans and achievements, and transparent community relationships. It also means resisting the urge to overinflate progress toward a goal like adding more diversity, increasing sustainable practices, or community engagement.

Much ink and column space have been devoted to the idea that Americans, and citizens around the globe, increasingly live in separate worlds with their own maxims, accepted tenets, and realities. To that end, according to recent Pew Research, half of U.S. adults obtain at least part of their information from social media.[7] Facebook is the most popular news source, with content and ads constantly skewed based on clicks and "likes" by each reader. The overwhelming majority of adults receive their news from digital devices, often through an app with AI in the background that delivers information that would be most likely preferred for that individual based upon their prior viewing habits.

As someone who has leveraged AI and data on many occasions to help facilitate positive behavior change, its power cannot be underestimated. For example, we recommend using AI for the retailers we advise to create personalization of messages, ads, special offers, and content to help customers save time and money by showing them more of what resonated with their interests. Also, ChatGPT has become the new best friend for copywriters, those digging through insights looking for trends, and students everywhere as it leverages AI to produce nearly instant suggestions.

Unfortunately, using these methods to deliver the news—facts about our climate, the Covid crisis, politics, and social unrest—skews the truth, and mistrust in news sources has never been higher. Not only does the method of obtaining news matter, but it seems our political affiliation also plays a large part in trust of media content. According to a Gallup survey, only 7% of Americans had a "great deal of trust" in the news. That is broken down further as 68% of Democrats, 31% of Independents, and 11% of Republicans have some level of trust for media content.

In business, you can't always control the variables in the political, environmental, and social landscape around you. But you can control the ethics and principles you hold and that shape your decisions and the way you communicate them both internally and externally. This can help you—and by extension, your team members, clients, and customers—feel more empowered in a time when truths, and power, seem exclusive, distorted, and qualified.

There is nothing new about large (or small) companies trying to cover up or manipulate the truth to meet their goals. What has changed, however, is workers' tolerance to put up with it. Improving your company's culture to align with truth and trust will ultimately function to attract and retain top talent. It starts by recognizing the contributing forces that create imbalance and finding ways to mitigate them. Overall well-being and engagement can only be achieved when workers trust that their organizations will balance their weight evenly on the teeter-totter to prevent them from crashing to the ground, getting hurt, and walking away from the game.

Chapter 3

THE BURNOUT CRISIS

According to my calculations, the fastest way to burnout is to try to be a "yes" person, when you know you're a "no" person.

—RICHIE NORTON

I knew from the first day we met at the neighborhood gym that I wanted to be friends with Maggie Ramirez. She had one of those magnetic personalities that lit up a room and instantly engaged the folks within it. We were both runners and worked in the tech sector, and often bounced ideas back and forth as we trotted on adjacent treadmills.

Maggie was an awe-inspiring leader at a Fortune 1000 company and served as chairperson of the company's new Diversity & Inclusion Council. Proud of her heritage, her parents immigrated to Alabama from Mexico before she was born. The first in her family to graduate from college and the eldest of four children, Maggie was comfortable taking charge and organizing the way for others. She was a natural leader and collaborator.

Sheer determination brought Maggie to Nashville to work for a high-growth retailer. She loved the fast pace and challenge of working in technology, especially helping her company understand how to better identify and serve their customers' needs. Maggie worked hard to get where she was, building her business network with focused effort over time. She was a critical part of the fabric that bound together the many technical capabilities demanded by her employer.

As many overachievers are, Maggie was a bit of a perfectionist at work. The first (and only) Latina director in the organization and the "face" of the diversity council, it was important to her to project a strong image. I was writing and speaking about burnout to employee resource groups (ERGs) and leader groups, so she and I often talked about the physical manifestations of too much stress over time. Yet, it was difficult for her to shed the pressures, both internally and externally generated, that amplified over time.

At the onset of Covid, the company's complex technical needs accelerated dramatically, which required unforeseen modifications at lightning speed. People who work in areas outside of technical support may not fully appreciate how quickly things changed during the pandemic lockdown period for retailers. **According to Pew Research, 40% of workers used technology in new ways, and McKinsey reported a seven-year accelerant of digital technologies for businesses during Covid.** Both accelerants required seemingly endless support from Maggie's team as years' worth of changes were conceived, implemented, and executed in rapid succession. Some people in her department didn't take a single day off for months, as they felt tremendous pressure to keep the business thriving.

During this time, unsurprisingly, the organization witnessed a significant drop in worker satisfaction scores on the work-life balance portion of its annual Employee Satisfaction Survey. Given the known propensity for acquiescence bias, or the tendency for workers to respond to surveys with answers they believe leaders want to hear rather than what they really feel, this major drop should have sent a strong signal to executive leadership that something needed to change, and fast.

Executives gave the green light to hire thousands of workers to handle the front lines in its retail stores. There were new safety requirements to be met and pickup services that needed additional help. At the home office, however, staff levels remained the same. Maggie and her team dutifully continued to work excessive hours.

By the end of 2020, they reached a point of fatigue so extreme, Maggie directly approached her boss to express concern about the team's burnout. She was way beyond trying to mitigate her team's frustrations and had become seriously worried about their health. One team member experienced a stress-related heart attack, while others were overall irritable and unusually inflated by situations they typically would have taken in stride. The work environment was like her mom's teakettle from childhood; she heard the water reaching its boiling point and was waiting for the whistle to blow.

Instead of being empathetic or offering a solution, her boss reminded Maggie that these were still uncertain times. It was important that she remained a good team player and set an example for others in her department. She shared her growing frustrations with me at the gym when we ran off the stress together after work. "He said to just keep going a little longer, but it's been over a year at this pace," she relayed. "He made me feel like I wasn't supporting the bigger organization by raising a red flag. It really hurts to be looked at that way after everything I'm sacrificing."

Maggie's words echoed ongoing sentiments of burnout that were exacerbated across the globe during the pandemic.

- 57% of workers felt they were required to give 24/7 accessibility to their employers (*Harvard Business Review*).
- Employees who feel unsupported by their managers are 70% more likely to experience burnout (Gallup).
- Employees struggling with burnout are 63% more likely to take a sick day and 23% more likely to visit the ER (Gallup).

Maggie's employer was a Wall Street retail darling. It experienced such a phenomenal rate of growth before the pandemic that when its growth slowed to only *twice* the national average, they braced for stock market reactions with layoffs. In the first quarter of 2021, Maggie was told to trim her already lean team by two workers. I remember how upset she was about it.

Working on sustainable business models for over 20 years, I understood the game of profits and perceptions, but Maggie and her team struggled to understand how the retailer could be so far ahead of its competition, yet still needed to cut back. The common corporate practice of dumping employees at the first sign of market tension was amplified in Maggie's case because her department was already seriously understaffed.

One of the team leaders she laid off, Aja, had moved to the U.S. a couple years earlier to support his family overseas. A few weeks after his dismissal, Aja took his own life. Maggie was haunted to the core that she may have played a role in his tragic decision.

More concerned than ever for her team's well-being and desperate for some support, Maggie approached her HR representative. Thanks to HR intervention, the tech department agreed to dive deeper into the work-life balance issue through a new, more specific survey. Team members were so frustrated by this time, they didn't sugarcoat their opinions. They spoke the raw truth. Feeling validated, Maggie shared that her team provided nearly a hundred detailed comments describing the factors that led to their negative work-life balance.

"Surely this new information, these personal cries for help, will change things," she said, since the company was now aware of precisely what needed to be fixed. Unfortunately, after Maggie shared the survey with her boss, she was abruptly instructed to "Lose the results." Maggie was numb with shock. "Lose the results?! What the hell am I supposed to do with that?"

She felt trapped as she vented to me, seeking comfort and political advice from an outsider. My heart ached for her, and I encouraged her to return to HR again or perhaps even consider changing employers.

Although Maggie felt disregarded by her boss, she also had a strong sense of duty to protect him and his peers from the new CEO. Maggie had a good relationship overall with her boss, and she enjoyed working with the Diversity & Inclusion Council, so she explained, "I don't feel like I have a choice. It's complicated."

After our conversation, Maggie didn't miss a beat at work. Nobody knew the depth of burnout and burden Maggie was carrying upon her shoulders and within her heart. At only 41 years old, just one month after our last conversation, my beautiful friend took her life on that crisp October morning to the horror and heartbreak of her family, friends, and neighbors.

What else could I have done? Why did she stay in that job despite feeling unheard? Was anyone at the retail giant going to tell executives the truth, or would they just position it as an unfortunate accident to keep covering their tracks? The sad truth is that Maggie's story is not an isolated event.

The Sad Reality

Millions of workers are experiencing similar challenges of burnout at work, but there is hope if leaders will recognize the signs, listen to their employees, and take positive action before it's too late. The conditions in recent years have created the "perfect storm" of uncertainty with each subsequent event increasing complexity and confusion.

Even before the Covid pandemic, a Gallup study reflected that 85% of Americans were unhappy at work. Deloitte released data stating that 77% of respondents said they experienced burnout at their current job. Job dissatisfaction was among several contributing factors.

In May 2019, burnout became such a predominant health concern that the World Health Organization (WHO) classified it as a syndrome, a condition resulting from chronic, unsuccessfully managed workplace stress. According to the WHO, burnout is characterized by three core symptoms:

- Feelings of energy depletion or exhaustion
- Increased mental distance from one's job, or feelings of negativism or cynicism related to one's job
- Reduced professional efficacy

An article by Robin Wright in *The New Yorker* speculated that the risks of depression and the type of post-traumatic stress disorder (PTSD) associated with war or natural disasters could likely have been even higher with the coronavirus pandemic because there was no definitive end date. According to the Centers for Disease Control and Prevention (CDC), 40.9% of adults reported experiencing at least one adverse mental or behavioral health condition related to conditions during Covid-19, including 26% of adults who reported symptoms of a trauma- or stress-related disorder.

The CDC further warned that individuals often experience an acute trauma response following a traumatic event. If this response lasts for an extended period, PTSD and other long-term mental health effects such as anxiety and depression are more likely to occur. Similarly, studies by Mental Health America showed the Covid-19 pandemic was itself a traumatizing event that was *also* coupled with other traumatic changes such as financial hardship, housing and food insecurity, death of loved ones, dramatic changes to work and schooling environments, and increased household stress that contributed to higher interpersonal domestic violence.

A pre-Covid survey by Indeed revealed that 43% of respondents were burned out at work. That percentage increased to 52% in 2021. The steady drumbeat of anxiety caused by the pandemic led to "systems overload" in many cases, resulting in repercussions never before experienced in the workplace. The pandemic essentially served as an accelerant to an already large problem; it added fuel to the flame of the burnout reality.

While the burnout syndrome is evaluated by its emotional impacts, the economic consequences are very real. Even before implications from the Covid pandemic, the Stanford University Graduate School of Business reported that **burnout costs $190 billion per year in healthcare expenses, in addition to 120,000 stress-attributed deaths in the U.S. alone.**[8] In fact, employee burnout became so bad in Japan that they invented a word for it, *karoshi*, or death from overwork.

Beyond healthcare, the web of interrelated costs to businesses is significant. Lack of engagement, higher turnover, the cost to replace talent, and low productivity are just a few of the ripple effects from burnout. In 2021, Gallup found that 80% of people are not engaged or are actively disengaged at work, partly as the result of stress. According to the *State of the Global Workplace 2021 Report*, lack of engagement costs the global economy $8.1 trillion in lost productivity each year. It translates directly into a lack of profitability, as teams with low engagement are **14% to 18% less productive** and have **turnover rates that are 18% to 43% greater** than highly engaged teams.

Gallup also found that burned-out employees cost 34% in salary because their productivity suffers when they are disengaged at work. But the cost is even higher when people quit! The Society for Human Resource Management (SHRM) reports that it costs a company **six to nine months of an employee's salary, on average,** to replace them. For an employee making $60,000 per year, that comes out to $30,000 to $45,000 in recruiting and training costs.

The Energetic Reality

Undeniably, burnout affects people both mentally and physically and negatively impacts the success of organizations. However, we don't often consider how burnout affects people energetically. When people feel fear, uncertainty, and other "low vibe" emotions, they can't engage or perform well at work. "Low vibe" forces—like a lack of security or support, feeling disconnected from your manager, being out of alignment with the corporate culture, and being in the wrong role—are forces that consume your "high vibe" positive energy.

According to Electrical4U, a reputable electrical engineering source, scientific evidence points to a fascinating idea: everything alive, including us humans, is composed of these tiny energy units known as "quanta." These quanta are like the building blocks of quantum physics, the science that delves into how matter and energy behave on the tiniest scale. This revelation has big implications—it suggests that matter and energy aren't two separate things, but rather different facets of the same intricate reality.

This concept becomes even more intriguing when we think about our own emotions and actions. You see, since we're essentially bundles of energy, the balance of positive and negative charges in our system matters a lot. **When negative influences outweigh the positive ones, it's like our mental and physical well-being hits a pause button.** It's a natural self-regulation process, sort of like a built-in defense mechanism.

The real magic happens when we find harmony between the pressures we face and the things that truly recharge us. It's like maintaining a stash of positive energy that we can tap into during the tough times and then replenish when the storm passes—a bit like a natural energy cycle. But, here's the twist: when negativity becomes a regular guest, lingering around without giving us a break, it can drain away our positive energy and put us at risk of burnout.

In essence, this insight from the world of quantum physics offers a fresh perspective on how our energy levels impact our well-being. Just like the delicate balance of charges determines the behavior of the tiniest particles found in all living things, finding our own equilibrium between positive and negative energy can shape our emotional and physical health.

In 2016, I experienced burnout to the degree that I quit my job as a successful executive, gave away most of my things, and lived in a remote jungle area of Costa Rica for three months. I was over an hour from paved roads and without TV, radio, or even cell phone connectivity. Fortunately, armed with a modern toolkit of books, yoga practice, and the ancient restorative power gleaned from plants and animals of all types, I was able to fully recover.

After some much-needed time away from working, I wrote a book called *Trust Your Animal Instincts: Recharge Your Life & Ignite Your Power*. It's about my personal journey to help others recognize burnout symptoms and causes, then tap into their own natural energy resources to achieve balance. The book's theme was that since energy is defined in physics as the capacity to do work, being drained of it means we lose our productivity (professionally and personally) and lack the sparks of curiosity, inspiration, and self-renewal. **Through an energetic lens, burnout is caused by doing things that consume, rather than recharge, our energetic resources.** At its worst, we are no longer able to feel consistently positive or recharge ourselves to a positive state.

Einstein said, "Energy cannot be created or destroyed; it can only be changed from one form to another." This truth is as relevant in boardrooms as it is in science labs. Picture your energy as a dynamic currency, always in motion but never lost. When you face those demanding challenges, it's like the intricate dance of energy—it reshapes the contours of your positivity. It's not just about psychology; it's about molding the essence of your potential.

Think about your emotions as the brushstrokes painted by this energy dance; it is the real-time tango between your positive and negative energies. This isn't just theoretical; it's the rhythm that fuels your decisive strides and shapes your strategic moves. Venturing further into the world of metaphysics, beyond the bustling meetings and spreadsheets, there's a profound undercurrent, your very nature of being. It's like the conductor guiding your symphony of actions, all rooted in this complex energy interplay.

So, as you navigate the intricate threads of business, remember that your energy isn't fleeting—it's transformative. **When challenges appear,**

they're not roadblocks; they're catalysts, shifting your energy's form and fueling your personal evolution. It's not just a psychological shift; it's the essence of your being, the dynamic force that guides your journey.

How does this show up in the workplace? Listen to the words spoken by your team, colleagues, or yourself to describe the current state. Words such as *isolated, cut off, out of sync, off track, spun up with anger, powerless, insulated, slowed, stopped, inhibited, forced, negative, limited, confined, restricted, frayed, fragmented,* and *out of balance* raise red flags that should not be ignored.

Traditionally, corporate culture encourages employees to fight through challenges rather than taking time to pause and process their impact. While many modern companies have flexible hours, work-from-home options, and meditation rooms, and encourage exercise breaks, it is still common in corporate culture to reward employees for working endless late nights and weekends. Employers call them out as dedicated team players and insinuate that those who draw boundaries around their personal time or strive for work-life balance are falling short of expectations.

Have you ever had a boss who complained constantly about having too much work but did nothing to improve their work-life balance? This behavior establishes a culture for burnout because the leader is actively demonstrating that *being miserably overworked is what leaders do and that it is expected.* It also sends an indirect signal that they are too busy to be approached, making it less likely for team members to ask for help. It creates a doom-loop cycle of frustration.

Inherent Natural Preferences

A contributing factor to increasing rates of burnout is when people perform tasks that are not in alignment with their natural strengths or interests, which creates dissonance and requires additional energy. Workers want to feel in sync with their team and their company cultures. Everyone has a natural preference toward activities or initiatives that energize them the most. For example, if someone is energized by inciting new ideas, but placed in a role that requires mostly conformity and risk aversion, over time, they will burn out. The same is true for those energized by finding efficiencies and producing stability. If required to work in a culture of constant change and variability, they could also be headed for burnout.

Those natural preferences for change extend from work life into personal life, as well. Being aware that I find change exciting helps me realize the feeling of being drained by spending long periods of time with someone who is completely the opposite—who enjoys looking for reasons why *not* to try something new. It also helps me appreciate the value of asking that very same person for advice when I'm considering an important decision to get a more balanced perspective.

Once we are *aware* of our energetic differences, it's easy to depersonalize the impact we're experiencing and value the opposing view as a complement to our energetic strengths. Most importantly, we must understand that every person's preference for getting into action is different and equally valuable.

Teams without risk-averse members may overspend, overpromise, and think short-term. Teams without risk-takers may look to the past for solutions and struggle with innovation. This variation in preferences for risk aversion and risk taking occurs within every individual as well, so success is about finding the right energetic balance within us and in our interaction with others.

Beating burnout will require more attention to human needs. Modern tools and science will be shared in the upcoming chapters of this book that are immediately available to help leaders seeking real solutions. Every organization has the opportunity to connect and align with its employees in a way that provides positivity instead of burnout. To start, companies need a more proactive, aware, and concentrated focus on what people need and want to prevent burnout and its related costs.

Despite the current state of technological change, imbalance, and burnout, there is hope for corporate culture through an active awareness of what energizes people and the importance of aligning their work with purpose. **Leaders have the ability to build a positive workplace culture that leverages each person's strengths. I call this active framework Human Dynamics** and will focus the remainder of this book on sharing it with you.

Part II

HUMAN DYNAMICS: POWER FROM THE SOURCE

Chapter 4
ENERGY: THE CATALYST FOR CHANGE

Everything is energy and that's all there is to it. Match the frequency of the reality you want, and you cannot help but get that reality. It can be no other way.

This is not philosophy. This is physics.

—ALBERT EINSTEIN

How can forward-thinking leaders catch up with change and leverage it for momentum and growth? How can we create a balanced culture that attracts and engages good workers? And how might we convert burnout to enthusiasm that is contagiously channeled among diverse teams, flowing back and forth unencumbered by friction? Further, what sparks people's curiosity, gets their blood pumping, and ignites them to take action?

These questions represent the direct challenges that executives, team leaders, and individual workers are confronted with in today's ever-changing reality. And yet, the answers to these questions boil down to one simple resource that's available to everyone at no cost and in every setting: energy.

Energy is defined as "the ability to do work," and energy is the basis for all life. It is the core power source that triggers all emotions, actions, and interactions needed to establish and maintain healthy people, teams, and organizations. It gives us the conduit to connect and provide input and feedback to foster individual and organizational growth.

My Lightbulb Moment:
Change = Connection + Direction

Not by coincidence, many activities and attributes within the workplace are described in energetic terms: burnout, connectivity, direction, output, production, transformation, transition, throughput, flow, power, vibe, and in-sync. The list goes on and on. And what is the universal symbol for a great idea? The lightbulb!

My own lightbulb moment came with the discovery of an ancient principle: the everything-ness of energy. Its universally applicable behavior throughout nature, bioscience, and electricity can be applied to predict human behavior, growth cycles, network development, and every other aspect of businesses. This was an ah-ha realization of epic proportions that cracked nature's code wide open! Imagine being able to predict how someone will react, increase the likelihood of success, improve adaptation, accelerate growth, bring balance, avoid burnout… all by simply emulating the laws of energy itself. It forever changed the way I viewed change management, human interaction, and leadership.

At the time, I was the innovation officer for our company, so I began developing a framework based on the ways energy naturally moves to help employees, clients, and customers adapt more quickly. This body of work, when applied to the ways people interact with each other, was christened Human Dynamics. It became a highly predictable way to leverage natural energy attributes and behaviors to increase the adoption of new ideas and propel the businesses forward.

This discovery, however, did not happen the way the phrase "lightbulb moment" might indicate. It came to us gradually over time, through trial and error, through studying and learning about electrical properties, through my work leading innovation (which required lots of behavior change), by pursuing certifications in holistic human biofield studies, and through testing various theories in companies of differing sizes and across multiple industries. For those not familiar with biofields, it is a field of radiating energy that is emitted from people ranging outward from their core and measuring approximately eight feet. While not visible to the untrained eye, biofields can be felt by physical touch or from an adjacent field, like when someone stands uncomfortably close. This groundwork, combined with a case of my own corporate burnout, led to the understanding that **adaptation and implementation of all business solutions boil down to simply managing human energy dynamics.**

It took me a little over a decade to put the pieces of the Human Dynamics puzzle together, which ultimately began with a whopping failure. It was the early 2000s when someone I respect said, "You may want to start looking for a new job." He scowled as he handed me the utilities expense reports for the 200 new Energy Star Certified homes our company, Actus Lend Lease, owned and operated.

Ray wasn't one to mince words. He served in four tours of duty on four different continents in military intelligence and then as a helicopter pilot instructor before he became a CPA specializing in asset management. I was SVP for sustainability and innovation (way before being "green" was cool). Our mission was to demonstrate significantly lower electricity costs among residents living in our brand-new certified green homes. We were eager to prove that homes with features like smart thermostats and high-quality windows would have lower energy consumption. But Ray's reports indicated precisely the opposite: electricity costs in our brand-new homes were using significantly **more** electricity than the Department of Energy's baseline for their house types. *This was not supposed to happen...*

Like most companies that roll out new initiatives, we assumed if we simply developed and shared educational information like energy saving tips with the residents, it would automatically lead to sustained behavioral change. Not so! Companies make this mistake all the time with HR policies, employee resource groups formation, sustainability program initiation, volunteer programs, standard operating procedures, and so on.

At Actus, we went so far as to develop an awareness and educational conservation program called SYNERGY—with a lofty promise to "Save Your Nation's Energy." Working with the local power provider, we sent lists of energy-saving suggestions to residents each month, assuming they just needed to learn how to conserve energy. We also conducted trainings for residents on the new smart devices in their homes and energy-saving techniques. While these educational initiatives raised awareness and goodwill, they did nothing to significantly and consistently change energy consumption behavior.

Disappointment and confusion swept through our ranks as we delved into the realm of behavior change. The shocker? We learned **the link between education alone and sustained behavior change in adults registers a big, fat zero.** Yes, zero! Further, for leaders orchestrating change from the top down, there are limits to the adoption of new ideas based on the amount of friction it causes in people's day-to-day lives.

For example, say your company wants to reduce expenses, rallying five dedicated minds on a committee to devise the approach. The team establishes 10 goals for the organization, then releases them to employees with great enthusiasm. Sounds like a solid plan from a typical committee, right? Well, here's the challenge: while workers may agree with the aspirations, a whopping 40% to 90% of human actions are habits. In essence, any new change in business becomes a balance of stopping the old habit and ushering in the new one.

The laws of energy dynamics show it's much harder to completely stop one behavior, then start a new one, than to simply tweak the direction of an existing behavior. Imagine how much more time and effort it would take to get anywhere if you had to stop a train completely, then restart it in a new direction instead of just merging it onto a different track. It's the same with implementing new ideas. What if the committee, instead, shared a few examples along with the overall goal—to lower expenses—then asked every employee to find just one way within the next two weeks to cut costs? This approach no longer requires workers to stop one habit and begin another; instead, it allows them to engage in a way that's meaningful to them and continue their momentum forward.

Getting new ideas into action is about flowing with the current that is already moving inside every person. **It's exponentially easier and faster to adapt when the change comes from personal choice versus a list of new requirements.** According to Rosabeth Moss Kanter, who writes about change management in the *Harvard Business Review*, "Resistance to change manifests itself in many ways, from foot-dragging and inertia to petty sabotage to outright rebellions. The best tool for leaders of change is to understand the predictable, universal sources of resistance in each situation and then strategize around them."

Still digging for answers, I came across Malcolm Gladwell's *The Tipping Point* and *Nudge* by Richard Thaler. Both authors clarified, in their own way, that igniting action requires two main things:

1. **Connecting** with each person directly (relevance)
2. Clear **direction** (1-2 actions at a time)

This simple formula hit home for me. I already knew from my electrical engineering studies that the most fundamental requirement for power and productivity to flow is connection to an energy source. The more direct a current, the less loss of power occurs as energy moves from one place to another and the greater its output.

At Actus, we needed to make the change more relevant and focus on giving **one clear direction** at a time that **connected personally** with residents. We had been diluting the residents' power by giving them too many options for action at once. What we thought was freedom of choice through education created overload that resulted in inertia. The dynamics of decentralization, where each person makes one small contribution, means potential for change generated is unlimited based upon the number of customers, residents, or participants multiplied by one thing at a time. In contrast, a committee's effectiveness can be limited by their personal sphere of influence to implement change.

What began as a failure with the Energy Star green homes served as an important catalyst to understanding what really moves programs and people forward, fast. Armed with what we learned about human behavior and inciting action through personal connection and clear direction, we partnered with Cornell University to create an educational energy program for another military community in New York. This time, we had just one specific focus—lighting—and so we concentrated our efforts on providing information about individual cost savings by turning off the lights when not in use.

There was obvious synergy when people were personally plugged in to the solution and turned on to a clear, shared direction.

We realized that we were on the cusp of a new way to accelerate adaptation by getting to the core of what drives each person, what really energizes them, and then finding ways to flow with that as simply as possible. This epiphany served as the foundation for reinventing the SYNERGY program across over 40,000 homes, and to our delight, the New York-based privatized housing community achieved a sustained energy savings of 12% (half of which could be attributed to technology changes; the other half was due to habit changes).

The everything-ness of energy, its governing principles, and its critical role in the business realm, particularly in terms of inciting action to adapt to change, was beginning to take a strong hold in my mind the more success we experienced by leveraging it.

Origin Frameworks

The Human Dynamics Framework is built upon ancient wisdom from the natural world. But I am not the first innovator who has attempted to identify what motivates people to change. Several touchpoint frameworks laid the groundwork for mine. Let's take a closer look at some of the more common and widely accepted bodies of work to see how they align with and are driven by the everything-ness of energy.

Maslow's Hierarchy of Needs

One of the best-known theories of human motivation is undoubtedly Maslow's Hierarchy of Needs. Abraham Maslow was a psychologist who studied positive human behaviors. His work was informed by the traditions of the Blackfoot indigenous people, whom he spent six weeks studying in the summer of 1938. His time among them, combined with a biographical analysis of 18 people he identified as "Self-Actualized," greatly influenced his "humanistic" psychological perspectives and assumptions. Ultimately, his research led to the development of his Hierarchy of Needs in 1943. The model was intended to explain human behavior based on their inherent and collective motivations.

MASLOW'S HIERARCHY OF NEEDS & MOTIVATION

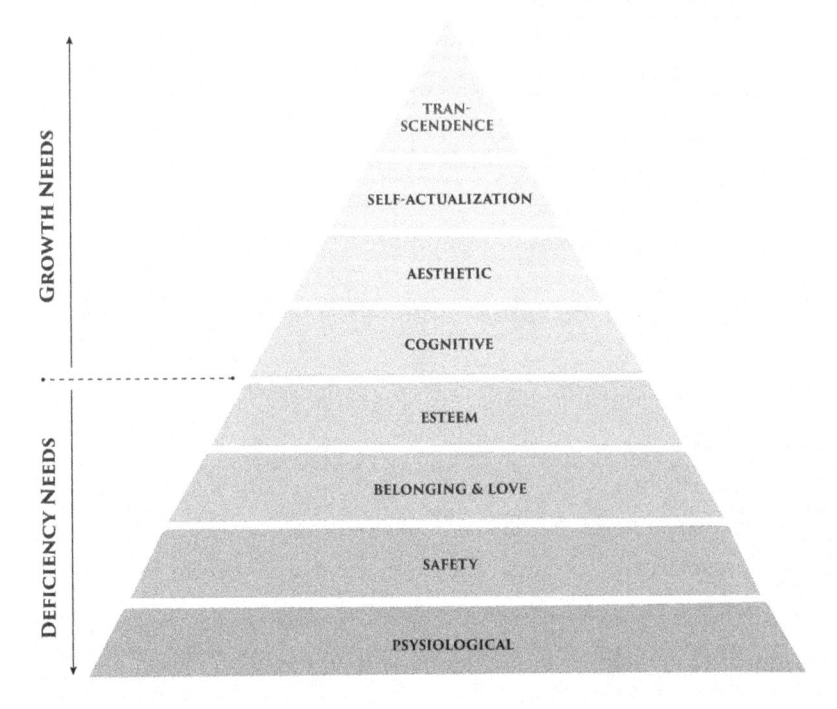

GROWTH NEEDS
- TRANSCENDENCE
- SELF-ACTUALIZATION
- AESTHETIC
- COGNITIVE

DEFICIENCY NEEDS
- ESTEEM
- BELONGING & LOVE
- SAFETY
- PSYSIOLOGICAL

Deficiency Needs: Upon achieving, motivation declines as energetic balance is achieved.

1. Physiological needs: air, food, drink, shelter, warmth, sex, sleep, etc.
2. Safety needs: protection from elements, security, order, law, stability, etc.
3. Belonging and love needs: friendship, intimacy, trust, acceptance, receiving and giving affection and love, affiliating/being part of a group (family, friends, work)
4. Esteem needs: Maslow classified these into two categories:
   - Esteem for oneself (dignity, achievement, mastery, independence)
   - The desire for reputation or respect from others (e.g., status, prestige)

Growth Needs: Upon achieving, motivation increases as one's energetic state grows more positive.

1. Cognitive needs: knowledge and understanding, curiosity, exploration, need for meaning and predictability
2. Aesthetic needs: appreciation and search for beauty, balance, form, etc.
3. Self-actualization: realizing personal potential, self-fulfillment, seeking personal growth and peak experiences
4. Transcendence: Motivated by values that transcend beyond the personal self (e.g., mystical experiences and certain experiences with nature, aesthetic experiences, sexual experiences, service to others, the pursuit of science, religious faith, etc.)[9]

All humans, regardless of culture, age, or preferences, have the desire and the potential to evolve into the higher tiers of the pyramid, but getting there is not as easy as it sounds. If someone's basic "deficiency" needs are not met, it's highly unlikely that their growth needs will be either. As a result, organizations must view deficiency needs as "table stakes," or the bare minimum requirement for engagement. Until workers achieve them, concepts that fall into the growth needs category such as innovation, design, and purpose will remain abstract and elusive.

Though Maslow's Hierarchy of Needs is almost a century old, his model has withstood the test of time. It is a strong visual representation of what humans need to go beyond merely existing to actually thrive and produce at a high level. Many similar models have been developed over time, with nearly identical overlap as they describe human needs and states of feeling or being.

Hawkins' Map of Consciousness

In the 1990s, Sir David R. Hawkins, M.D, Ph.D, developed the Map of Consciousness. Like Maslow's hierarchy, the framework was a way to understand human psychology and its spectrum of well-being. A renowned psychiatrist, physician, researcher, and lecturer, Hawkins developed a framework to categorize human emotions, thoughts, and levels of consciousness based on a numerical scale ranging from 0 to 1,000, with higher values indicating higher levels of consciousness. The framework includes states of being that range from shame and guilt to peace and enlightenment.

Hawkins assigned an energetic log or numeric score as a quantifier for the differences in feeling high-vibe states versus low-vibe states. Note the similarities to Maslow's framework with survival needs at the bottom and spiritual actualization at its top.

MAP OF CONSCIOUSNESS
DEVELOPED BY DAVID R. HAWKINS

Name of Level	Energetic Log	Predominant Emotional State	View of Life	God-view	Process
Enlightenment	700–1000	Ineffable	Is	Self	Pure Consciousness
Peace	600	Bliss	Perfect	All-Being	Illumination
Joy	540	Serenity	Complete	One	Transfiguration
Love	500	Reverence	Benign	Loving	Revelation
Reason	400	Understanding	Meaningful	Wise	Abstraction
Acceptance	350	Forgiveness	Harmonious	Merciful	Transcendence
Willingness	310	Optimism	Hopeful	Inspiring	Intension
Neutrality	250	Trust	Satisfactory	Enabling	Release
Courage	200	Affirmation	Feasible	Permitting	Empowerment
Pride	175	Scorn	Demanding	Indifferent	Inflation
Anger	150	Hate	Antagonistic	Vengeful	Aggression
Desire	125	Craving	Disappointing	Denying	Enslavement
Fear	100	Anxiety	Frightening	Punitive	Withdrawal
Grief	75	Regret	Tragic	Disdainful	Despondency
Apathy	50	Despair	Hopeless	Condemning	Abdication
Guilt	30	Blame	Evil	Vindictive	Destruction
Shame	20	Humiliation	Miserable	Despising	Elimination

Spiritual Paradigm (Enlightenment–Love)
Reason & Integrity (Reason–Courage)
Survival Paradigm (Pride–Shame)

Source: Hawkins, David R. M.D, Ph.D, *The Map of Consciousness Explained: A Proven Energy Scale to Actualize Your Ultimate Potential.* Hay House Inc., October 20, 2020.

Project Aristotle

Project Aristotle was a two-year study conducted in 2012 by Google to better understand the factors that contributed to their most successful teams. Interestingly, their conclusions coincide directly with Maslow's and Hawkins' earlier findings. Led by a selection of statisticians, engineers, sociologists, and behavioral psychologists, the researchers studied 180 diverse teams, searching for the magical "recipe"

PROJECT ARISTOTLE

- IMPACT & RESULTS
- MEANING & PURPOSE
- STRUCTURE & CLARITY
- DEPENDABILITY
- PSYCHOLOGICAL SAFETY

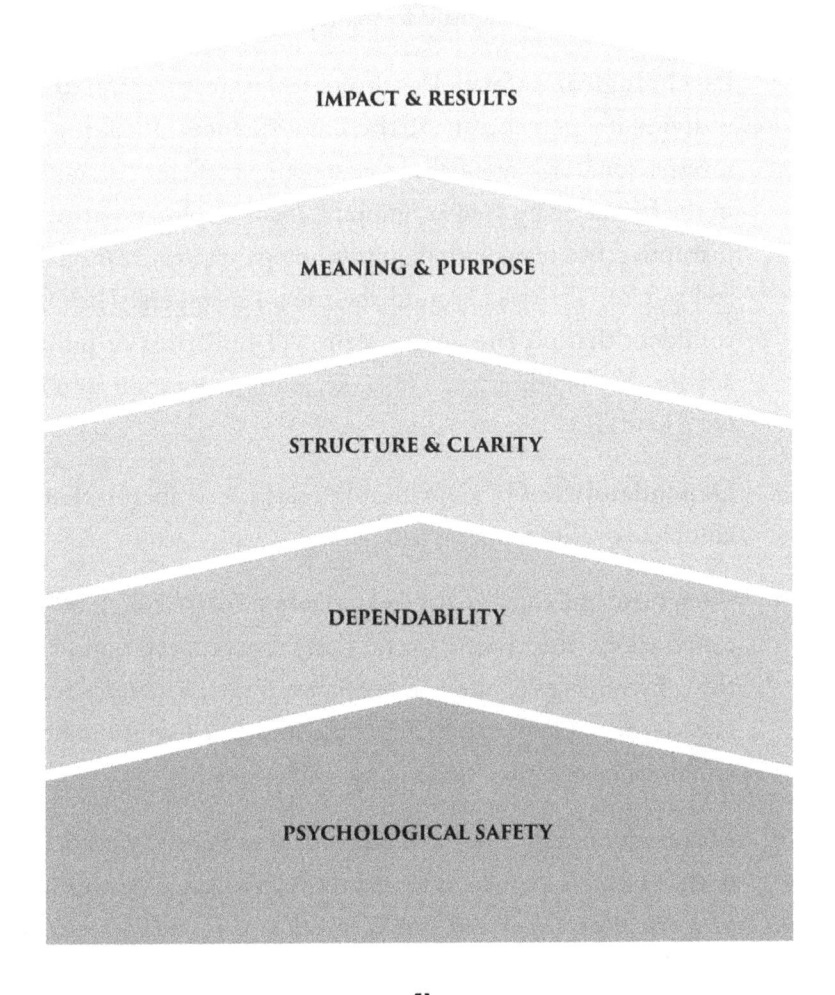

of team effectiveness. These results support the critical need for trust as one of the strongest pillars of human connection.[10]

Researchers discovered that what really mattered to people was how the team worked together—in other words, their *dynamics*.

The factors that Google's high-performance teams had in common are:

- **Psychological safety:** Psychological safety refers to an individual's perception of the consequences of taking an interpersonal risk or a belief that a team is safe for risk taking in the face of being seen as ignorant, incompetent, negative, or disruptive. In a team with high psychological safety, teammates feel safe to take risks around their team members. They feel confident that no one on the team will embarrass or punish anyone else for admitting a mistake, asking a question, or offering a new idea.

- **Dependability:** On dependable teams, members reliably complete quality work on time.

- **Structure and clarity:** An individual's understanding of job expectations, the process for fulfilling these expectations, and the consequences of one's performance are important for team effectiveness. Goals can be set at the individual or group level, and must be specific, challenging, and attainable.

- **Meaning:** Finding a sense of purpose in either the work itself or the output is important for team effectiveness. The meaning of work is personal and can vary, such as financial security, supporting family, helping the team succeed, or self-expression for each person.

- **Impact:** The results of one's work, the subjective judgment that your work is making a difference, is important for teams. Seeing that one's work is contributing to the organization's goals can help reveal impact.

In a comprehensive article about the project for the *New York Times*, bestselling author of *The Power of Habit*, Charles Duhigg, observed:

> "What Project Aristotle has taught people within Google is that no one wants to put on a 'work face' when they get to the office. No one wants to leave part of their personality and inner life at home. But to be fully present at work, to feel 'psychologically safe,' we must know that we can be free enough, sometimes, to share the things that scare us without fear of recriminations. We must be able to talk about what is messy or sad, to have hard conversations with colleagues who are driving us crazy. We can't be focused just on efficiency. Rather, when we start the morning by collaborating with a team of engineers and then send emails to our marketing colleagues and then jump on a conference call, we want to know that those people really hear us. We want to know that work is more than just labor."

Project Aristotle's five success factors line up nicely with the human motivations as depicted in Maslow's Hierarchy of Needs. The study also aligns with findings from Hawkins' Map of Consciousness, wherein higher vibe states, like engagement and harmony, contain the potential for higher productivity and innovative outcomes.

While Project Aristotle was a two-year initiative undertaken by one of the world's leading tech giants with unlimited financial and intellectual resources, the findings boil down to meeting basic human needs.

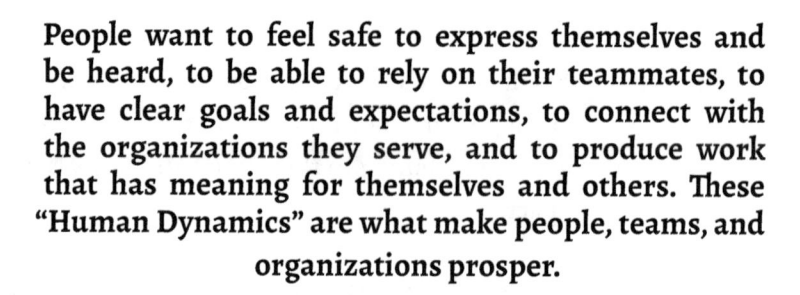

> **People want to feel safe to express themselves and be heard, to be able to rely on their teammates, to have clear goals and expectations, to connect with the organizations they serve, and to produce work that has meaning for themselves and others. These "Human Dynamics" are what make people, teams, and organizations prosper.**

If Google, the corporate data king, can create an environment that recognizes and celebrates the dynamics that make humans thrive, surely there's hope for the rest of us.

Ancient Human Energy Centers

Years ago, I went deep into the study of holistic medical practices that balance human biofields and was exposed, for the first time, to the seven primary energy centers, called chakras, within the body. The chakra terminology originated in India from the yoga *Upanishads*—dating back between 700 and 500 BC. Expansions came from a number of sources over the years and were translated for Western distribution by Arthur Avalon, in his book *The Serpent Power*, published in 1919.

The program I participated in studied 32 holistic techniques originating from the Far East, the Americas with the indigenous Hopi, as well as modern NASA physicists and Western physicians. Perhaps you have heard of Reiki, an example of one popular human energy balancing method widely practiced today.

The chakra model stems from an ancient map of energy flow throughout the body, and the word *chakra* means "wheel" because it is a spinning disk of energy. According to yogis, there are three major channels that flow along the spine. One, the Sushumna, flows up the center of the spine, with the other two, Ida and Pingala, coiled around it.

Interestingly, the caduceus symbol used in modern times for Western medicine echoes striking similarity to the ancient yogic channels of energy flow along the spine with two snakes coiled around a central staff, coming together in seven points as their faces meet at the top. The symbol is attributed to Hermes Trismegistus, who was written about in Greek and Egyptian texts as a great figure. He wrote about the interrelationship between the material and the divine realms—just another confirmation of our energy everything-ness.

At their seven intersection points, primary centers of power are created that progress from the base of the spine up to the top of the head, each spinning with higher frequencies as they progress.

The assertion is since we are made of energy, and we have these centralized centers of energy throughout the body, we can connect, increase, and balance them to maintain mental and physical wellness.

---

1. **Root Chakra:** Located at the base of your spine, the root chakra is responsible for your sense of security, survival, safety, stability, and comfort.

2. **Sacral Chakra:** Located just below the bellybutton and above the pubic bone, it's the center for sexuality, pleasure, and creativity.

3. **Solar Plexus Chakra:** Located in the lower abdomen, it is responsible for self-confidence, identity, and self-esteem.

4. **Heart Chakra:** Located in the middle of your chest, it is responsible for the ability to love and connect with yourself and others.

5. **Throat Chakra:** Located at the base of the throat, it is connected to communication and speaking truth.

6. **Third Eye Chakra:** Located in the center of your brows and forehead, this chakra is responsible for strong intuition and is also connected to wisdom and imagination.

7. **Crown Chakra:** Located at the top of your head, it is considered the chakra of enlightenment and represents our connection to our purpose and spirituality.

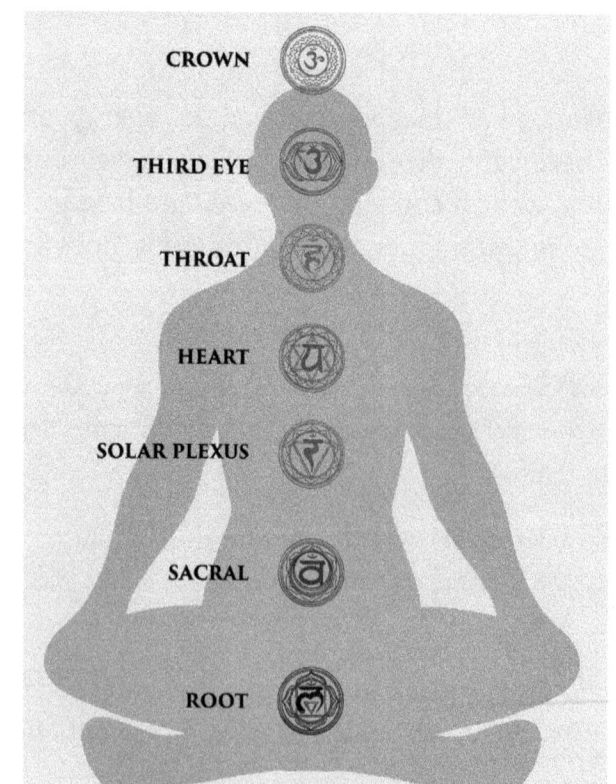

ANCIENT CHAKRA ENERGY CENTERS

Placing your palms on different energy centers creates an electrical circuit so energy can flow and become balanced throughout. As someone professionally trained and educated in the laws of electrical science, the ancient practices of energy balancing for people just seemed logical to me. When you understand the laws of energy, it demystifies a lot of mysticism. I remember after completing the first class thinking, "Well of *course* this works—connecting chakras is no different than jump-starting the battery in a car."

The Common Roots of Behavior

Upon learning about the Chakra System, it became apparent how the naturally occurring energetic system informs and impacts all other psychological and physiological systems, solidifying my understanding of Human Dynamics and their importance in the business realm. I was already well familiar with Maslow's Hierarchy of Needs as it related to worker engagement and performance and had studied David Hawkins' work before writing my first book on recharging from burnout. Google's Project Aristotle further supported Maslow and Hawkins' findings, and all three of the frameworks align to the ancient energy centers.

The following table demonstrates how all three of these well-known behavior frameworks align precisely with the human energy center mapping documented thousands of years ago.

In other words, there are many models of behavior and many more will be written in the future, but there is only one truth underpinning them all—energy is the basis of all life, growth, feeling, and action.

The visual parallels between the four models were too compelling to ignore, and I wasn't the first to see similarities. Anodea Judith, Ph.D, makes eloquent comparisons between Maslow's needs and the functions empowered by each major chakra in her book *Eastern Body, Western Mind: Psychology and the Chakra System as a Path to the Self.*

My additional assertion is that the similarities don't stop there. I believe the Chakra System aligns with any well-tested scientific model applied to human and organizational movement because it is based on how we get into action at our core level. Everything else, our emotions and our actions, are based upon that foundation.

Ancient Energy Centers & Related Functions (700 – 500 BC)	Maslow's Hierarchy of Needs (1943)	Hawkins' Map of Consciousness (1995)	Google Aristotle Project Findings (2012)
Crown: enlightenment and represents our connection to our purpose and spirituality	Self-actualization & Transcendence	Enlightenment	Impact & Results
Third Eye: intuition and is also connected to wisdom and imagination	Aesthetic	Peace & Joy	Meaning & Purpose
Throat: communication and speaking truth	Cognitive	Reason	Structure & Clarity
Heart: ability to love and connect with yourself and others	Belonging & Love	Love, Acceptance, Willingness, & Neutrality	Dependability
Solar Plexus: self-confidence, identity, and self-esteem	Esteem	Courage & Pride	Psychological Safety
Sacral: center for sexuality, pleasure	Safety	Anger, Desire, & Fear	
Root: sense of security, survival, safety, stability, and comfort	Physiological	Grief, Apathy, Guilt, & Shame	

Human Dynamics: The Energetic Imperative Starts at the Core

Human Dynamics is the framework that is needed now to cope with urgent workplace challenges, to connect with the people who drive our organizational initiatives, and to ensure we are tapping into their unique, individual power. It is based on the scientific principles of energy throughout the natural world and the laws of physics, which we'll explore through the context of business in the coming pages.

When leaders understand energy dynamics, first on the human level and then on the team and organizational levels, they can leverage its principles to connect, engage, direct, and inspire individuals in a highly predictable way, as my team and I discovered in New York with our lighting initiative.

At our core, there is no difference between any living creature, human, or business. Energy is the great equalizer! Renowned Nobel Prize-winning physicist Niels Bohr validated that all matter and living things are made up simply of whirling energy at the subatomic level, while famous physicist Erwin Schrödinger asserted, "I insist upon the view that all is waves."

Schrödinger's remark offers an important context. Our state of being oscillates between positive and negative energy; it ripples back and forth like the waves of the sea, giving us the capacity to generate greater positivity (high-vibration activities) or negativity (low-vibration activities). People's reactions to positive or negative changes and internal or external influences can be easily anticipated, based on the ways energy behaves.

This concept is where my work differs from other corporate innovators because my Human Dynamics framework focuses on ontology, or the way a person is being, and the central question: How do we provide the right environment for people to vibrate at a higher frequency? To enact change and to inspire high-level functionality, we must focus on people's energetic state (which dictates how we are acting and how we connect), tap into what motivates us, and move forward.

Getting to the core of what drives each person, what really energizes them, and finding ways to flow with that as simply as possible are critical steps to accelerate adaptation and change behaviors.

It's about how we leave inertia behind and build momentum in a forward direction, leveraging energy as the source of all thoughts and subsequent behaviors, going a layer beneath the manifestation of emotion, as indicated in the simplified diagram that follows.

ENERGY: THE TRUE HUMAN RESOURCE

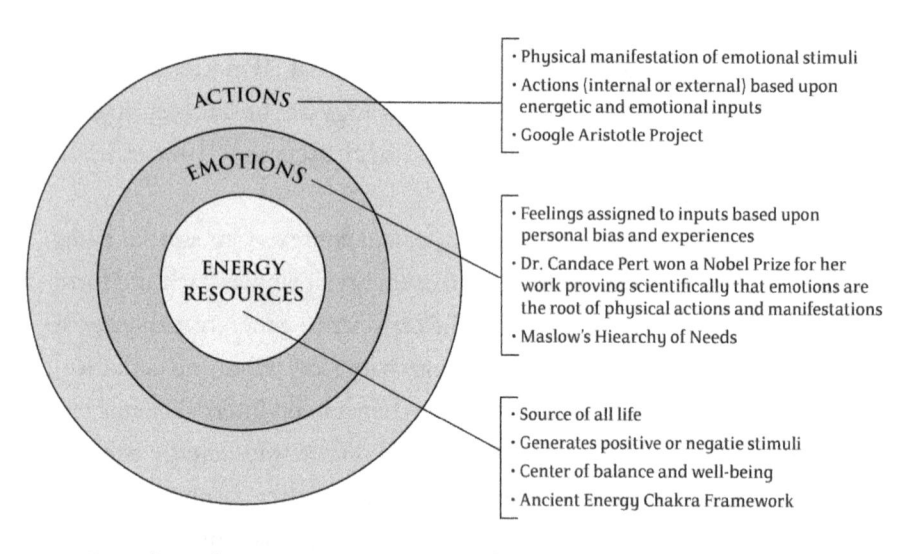

ACTIONS
- Physical manifestation of emotional stimuli
- Actions (internal or external) based upon energetic and emotional inputs
- Google Aristotle Project

EMOTIONS
- Feelings assigned to inputs based upon personal bias and experiences
- Dr. Candace Pert won a Nobel Prize for her work proving scientifically that emotions are the root of physical actions and manifestations
- Maslow's Hiearchy of Needs

ENERGY RESOURCES
- Source of all life
- Generates positive or negatie stimuli
- Center of balance and well-being
- Ancient Energy Chakra Framework

Based on the positive or negative forces received, the middle ring is where we tune in to it and assign emotions, based on our life experiences and related bias. The outside layer represents the physical reactions to the prior two inputs. It is where traditional businesses spend time trying to solve problems because most business leaders are focusing on the symptoms and results as separate systems (e.g., work and life) rather than the root causes of imbalance that flow outward from the current underneath.

The core, or shared source of all energy, is the catalyst that ignites emotions and reactions. Consider the core in nature; it is the area containing the power to continue the next life cycle. Future success depends upon what is generated at that base level, like seeds in a fruit, the sun within our solar system, eggs in a mother's belly, and so on.

I heard Bishop T.D. Jakes give an engaging talk about motivation using the core of an apple as a lens to better understand ourselves. He explained there's an orchard inside every apple. Future trees that will bear fruit are represented in each seed and may be accessed in a couple of ways:

- Cut right to the core.
  In people, potential energy can also be recognized, directly accessed, and leveraged from the core.

- Wait for the apple to rot, which will eventually reveal the same seeds.
  In people, this is the equivalent of burnout, depression, or fear of unpleasant consequences, which often serves as the transformational catalyst to change or adapt.

Bishop Jakes' story reinforces the concept that our individual core energy fuels growth and success, just as energy fuels the natural world. Since energy is the core of all manifested emotions or actions in the workplace, understanding and managing it is the only way to resolve the "symptoms" in the outer two layers of emotions and reactions (see previous diagram).

Further, **people and ideas are the seeds within a living organization.** Imagine the increased probability of success for the seeds that have been thoughtfully planted the optimal distance apart, in fertile soil, and with an abundance of water and light, versus the ones that drop to whatever part of the ground where the fruit fell and rotted. Similarly, the potential within every employee is optimized when the right business culture intentionally fosters their growth. Leaders can recognize the value and potential within each person, designing a culture that will create the highest possibility for strong growth.

The most successful culture is one where diverse ideas and people are consciously provided with space for expansion, nutrients like training and development, and encouragement for engagement. Recognizing ideas as potential resources also allows regeneration of product life cycles, maximizing and extending resilience over time.

Operation: Good Vibrations

Einstein said, "Everything in life is a vibration." Whether leveraging the findings of Google's Project Aristotle, Hawkins' Map of Consciousness, the Upanishads' Chakra System, or even Human Dynamics, the basis of every single model boils down to the energetic cause and effects of positive and negative vibes. Each of the zones in all the models has an electrical frequency attached to it, and according to the laws of physics, people must be energized at a certain level to resonate with higher frequencies.

If we're operating from a state of lower frequency, where much of the workforce is right now, concerned about survival and safety, feeling depleted and burned out, it's scientifically impossible to engage in higher frequency outputs such as innovation, imagination, creativity, balance, meaning, or purpose. The good news is that it is possible for anyone to increase their frequency and become receptive to engaging in those high-vibe, fulfilling types of pursuits.

ENERGY: THE CATALYST FOR CHANGE

This concept—the everything-ness of energy and how it affects all that we do in our personal and professional lives—is at the core of understanding and applying the Human Dynamics framework.

This phenomenon can be explained scientifically by defining "low vibe" versus "high vibe" states. In the 1970s, physician and researcher Dr. Joseph Puelo identified six vibrational frequencies used for sound therapy to heal human chakras. Vibrational medicine is providing breakthroughs in our modern universities today, but it has been used since ancient times within the practices of Western Christians, the Gregorian Monks, and Eastern Indian religions. Dr. Puelo's frequencies identify the corresponding vibe for each healthy chakra.

For math nerds, Dr. Puelo's frequencies apply to all other models, too, like Maslow's Deficiency Needs range from 396 Hz to 639 Hz, while Growth Needs range from 741 Hz to 963 Hz. The science is clear: Operating in an environment of fear and uncertainty makes it impossible to be as productive as when operating from within a more fulfilling environment.

Evidence in our modern world points to the truth that there is no longer a clear separation between work and life. Instead, all things are interconnected at a visceral level at which well-being is achieved through balancing the things that drain and ignite our energy. To do meaningful work with this context in mind, we must begin with an understanding of energetic balance at its core.

Anyone can see physical energy at work. We operate our phones, our automobiles, and our physical bodies as we move around space. It's easy to see the principles of cause-and-effect of physical obstacles and forces—your phone ran out of battery, your car must navigate around that obstacle in the road, you move from one place to another.

It's more difficult, however, to detect unseen energy, but you can certainly *feel* it. It's that sapping feeling you get in your gut when someone calls who always brings you down. It's the positive charge you get when someone gives you affirmation. It's the headache, neck ache, or backache you feel after doing things or being around others that don't resonate with your vibe. There's a common energetic euphemism for this effect, when someone or something is deemed a "pain in the neck."

Though it's not common practice to think of people as energetic beings, vibing at various frequencies, the simplicity of this perspective is fully backed by scientific laws, such as these, that provide a basis for understanding the imbalance we are feeling today.

- Since everything is made of energy itself, it is the basis for all action and reaction in the universe. All human actions and reactions follow the laws of physics and motion. Diagnosing and designing solutions at the root *cause* of imbalance, therefore, is a more direct approach to change behavior.

- The First Law of Thermodynamics states that energy can be changed from one form to another, but it cannot be created or destroyed. In other words, the total amount of energy and matter in the Universe remains constant, merely changing from one form to another with transition. What this means for people is that we already have a supply, a resource of unlimited, renewable, positive energy that is in constant transition based upon the stimuli and forces around us.

- The Second Law of Thermodynamics explains what's happening to our bodies and minds when we are around things that boost positivity (higher vibrations) or drain us with negativity (lower vibrations). It deals with the transfer of heat and energy inter-conversions. One simple statement of the law is that heat always moves from hotter objects to colder objects. This law explains why we feel drained, off-balance, or burned out after enduring too many negative experiences without balancing them with offsetting positive experiences. It also ties nicely to Maslow's deficit and growth theories.

These three foundational pillars describe precisely what's happening to your positive energy when surrounded by negative energy. Imagine you're feeling positive, in a high-vibe state, like a warm cup of your favorite coffee. Then, you must interact with a person or perform a task that requires operating at a lower vibe. Doing things or being around people who are operating at a lower vibration feels like a giant ice cube plopped into your biofield beverage.

The laws of physics dictate that your warm, high vibe *will* be lowered to create balance, compensating for the cold, lower energy vibe, unless you protect it. Sometimes these people are deemed "emotional vampires," and this phenomenon can also be seen when workers become "frozen in fear." Their inertia comes from a lack of ability to clearly see positive alternatives; therefore, they must sit, frozen, in their current state.

Energy is at the core of all our modern workplace ills, because when energy is zapped, it requires a charge to reboot. A leader's job is to recognize that without a strong existing baseline vibe, it cannot be built upon at creative and strategic levels. They're just trying to get juice out of a dead battery. It simply will not work. The battery needs a jump, as does our workforce, and to administer the charge correctly, we need to create connection to the power source. Connectivity is the beginning of the energy chain. It is the spark that ignites action, followed by momentum and acceleration.

Tapping the Current: Switch4Good

Connecting to the hidden positive power potential within every person and leveraging it exposes an ongoing, ever-present current. It's the current that hums in the heartbeat of every human being and at the core of every organization. It is the source of power driving adaptation, production, innovation, and change.

In 2012, I had the opportunity to experiment with the Human Dynamics framework and all I'd learned about behavior, energy, electricity, and accelerating change on an organizational level. I was serving as SVP of innovation and sustainability for a London-based company that owned a vast global portfolio of commercial buildings and mixed-use communities.

My role was to launch an energy conservation program called Switch4Good. Residents could opt in or out of the program if they were interested in saving on energy bills, and they were in control of how they were contacted (by text, call, email, social media app, etc.) so it was relevant to them and they were in charge of how to interact. Because they already valued the idea of saving energy and were involved in selecting the information delivery process, we hoped to fuel that momentum.

Switch4Good utilized data from smart meters in homes on six Navy installations in the Southeast to identify specific energy use habits. For example, if energy spiked at 6:30 a.m., Monday through Friday, we assumed residents were getting ready for work or school. We then generated a very specific habit nudge with clear direction, like "Try taking a shorter shower." Once the tip was given, an AI algorithm noted if consumption went up or down and determined whether to send more tips or discontinue that type of tip.[11]

The Switch4Good program was a huge success. We reduced energy consumption on average by 15%. People were changing and adapting because we learned how to *accelerate* their needs and wants rather than ask them to stop current actions and move into a new direction. It takes a lot less energy to add momentum than to start over from a point of inertia (like a train gliding onto a new track versus stopping and restarting in a new direction).

The results were so successful that we completely redesigned the company's innovation program. The Innovation Lab, or "i-Lab," was the first major corporate culture program where I deliberately applied the Human Dynamics Framework to accelerate the generation, adoption, and implementation of new ideas, but not until we learned some lessons the hard way.

Originally, and prior to the Switch4Good program, executives volunteered "up and coming" leaders from their international professional development program. They asked me to make it about innovation—something tangible we could use to improve the business. The first year, we allowed participants to create their own, innovative idea. I led them through the process of evaluating the concept and proposing solutions, complete with predictive financial models. The team's proposed solution would save the organization millions of dollars in lifecycle costs, but it was not in the current budget and there was no available staff to implement the work. So, while the executive team praised the group of young leaders for their ingenuity, the idea was never implemented.

You can imagine the repercussions. We asked the participants to create an idea that saved money. They delivered, and then the company did not implement it. Consequentially, the participants felt unheard, ignored, and devalued. They couldn't believe that short-term budget and staffing barriers would block such a high-impact, long-term solution. These feelings evoked personal disconnection, making employee engagement for innovation worse instead of better.

The following year, after the success of the Switch4Good program, I received permission to completely rewire our i-Lab. This was when we implemented the Human Dynamics Framework and put what we'd learned to the test. I started by interviewing board and executive team members to glean their top priorities. Then, I collected the entire "wish list" and asked them to sign off on the top one, which was to determine how smart technology could best be deployed in student housing. This gave me the direction they wanted to go, so any ideas generated would add momentum, instead of creating an equal and opposite effect due to competing actions.

Instead of "volunteering" people for participation in the program, we made it an application only process for two reasons:

1. If people took the trouble to fill out an application and sign on for the additional workload beyond their day-to-day activities, they were sincerely interested in learning through innovation. This meant we already had a point of connection from which to engage and build momentum.

2. By taking applications, we obtained team members who, until then, were "unseen stars." Introverts came out of the woodwork, so we found potential energy we didn't know existed.

At the time, we were involved in an active proposal to build and manage the student accommodations at a university in southern England. Our i-Lab team, among many other things, contacted the university to understand more about smart tech and student engagement through new technologies. We described our innovation process and how we had an active learning environment, with student housing technologies being the current focus.

It was a win-win-win! I-Lab participants were engaged and connected because the process of innovation energized them. Leaders were pleased because the ideas generated were immediately applicable and accelerated their current strategic plan. And the university recalled their request for proposal, and then re-released it to potential builders. They modified it to add points in the scoring rubric for having an active, ongoing innovation program.

Our company was selected for the project, which was the largest university project by bond issue in the history of the UK. The i-Lab changes reinforced the importance of identifying current direction, connecting, and adding momentum where the previous program created equal and opposite reactions that canceled out progress and created the perception of a corporate culture that was not listening.

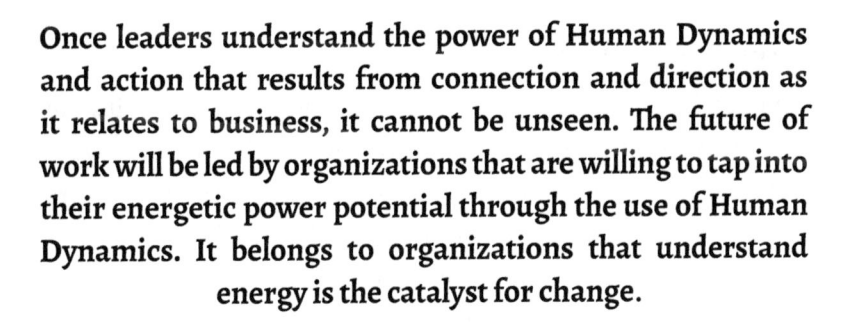

Once leaders understand the power of Human Dynamics and action that results from connection and direction as it relates to business, it cannot be unseen. The future of work will be led by organizations that are willing to tap into their energetic power potential through the use of Human Dynamics. It belongs to organizations that understand energy is the catalyst for change.

Chapter 5

THE GROWTH CURVE: NATURE'S NAVIGATIONAL TOOL

*In three words I can sum up everything
I've learned about life: it goes on.*

—ROBERT FROST

Improvement: It's Cyclical

If anybody knows about speed, agility, and cyclical training, it's Steve Keith. "Most people just call me 'Coach,'" he said the first time we met. Recently retired from Vanderbilt University where he coached track and cross-country, he was named the South Region Coach of the Year three times while there. And before Vanderbilt, he coached at Emory, Georgia Tech, UTEP, and Alabama.

We had been neighbors in downtown Nashville for over three years, long enough for me to appreciate Coach Keith's former and current accomplishments in the competitive world of sports. An intense cyclist year-round, he travels to exotic locations like Greece to ride hundreds of miles in the mountains and often outpaces men half his age on weekend rides in Tennessee. While we often saw each other heading out or returning home from cycling along the Greenway, riding together would not be an option. I could never keep up! I'm a leisure cyclist, riding more for respite than racing or record-breaking.

THE GROWTH CURVE: NATURE'S NAVIGATIONAL TOOL

Because his career and passions were devoted to training and endurance, both physical and mental, I was interested in talking to Coach about the bioscience of human adaptation, action, and motivation. My theory is that everything inside and outside of business follows the same life cycles found in nature, and I wanted to hear his perspective. Did he think that the human growth trajectory, when training for athletic competition, is similar to a company's growth trajectory as it matured? A humble introvert, Coach agreed to share his views in exchange for a few adult beverages at the pub across the street.

We talked about strategies and preparation for race days as well as the cycles of training in between major competitions. Coach Keith borrowed the pen and paper I was using to take notes and drew a series of escalating curves across the page, each one incrementally higher than the next. "If you want to grow, get stronger, and win, you have to build up through a process like this," he said, "then drop back to recover, and start with something different on a next curve."

He explained that training requires the cyclical introduction of new methods to catalyze and challenge the body in different ways based upon a particular goal. "Athletes go through a process of adaptation to the new workouts. As their body masters each technique, they need to switch up the method to use different muscle groups; otherwise, they'll overly stress the body. The process repeats over and over again, and recovery is necessary in between cycles."

Coach Keith planned and timed his athletes' training cycles to achieve a point of peak performance on race days, following a curve approach. Based on his successful "track" record, the method certainly seems to work.

Phases of Growth: They're Predictable

No matter how fast you run or how much you train, you'll never catch up with change. The nature of change is to disrupt the status quo, or evolve, so trying to control it wastes unnecessary energy. And critically, **despite the popularity of the phrase "change management" in** organizational design, change is not something to be "managed." It is an inevitable force that can be anticipated based on the laws of physics. (From its Greek origins, the word "physics" literally means "nature.")

Leaders must learn to *flow with* change, to adapt the way they look at it, and recognize its predictable patterns throughout all living things, including businesses.

Growth itself is defined as *an increase* that may be measured in many ways. For companies, traditional measures include things such as profits, value, size, return, and market share. Growth is powered by energy as it follows the same patterns, referred to as the "Constructal Law" of nature, throughout all bioscience and business. The economy, population growth, river erosion, and even our own bodies are designed with the same predictability.

Anything that grows has a natural biorhythm and follows the same Growth Curve with four distinct phases: birth, growth, maturity, and decline. When plotted on a graph, this pattern of growth has an "S" shape, so the Growth Curve is often referred to as an S-curve. The terms are used interchangeably.

- **Birth:** Birth is the phase for startup organizations, appearing at the inception of an S-curve, where new ideas and businesses are generated. This can occur in a number of ways based upon the circumstances. At times, opportunities arise from looking where everyone else is looking yet seeing what no one else is seeing.

- **Growth:** The growth phase occurs when organizations with established and tested ideas are expanding. The skills and investments to design a successful future for the new idea must be available and applied intelligently to properly nourish growth. These design skills are not common in organizations, and when present may be neglected. According to Harvard Business School professor Clayton Christensen, over 30,000 new products are introduced every year, and 95% of them fail. Most new initiatives fail because what they have in inspiration they lack in sound, sustainable foundations for startup development, and growth.

- **Maturity:** Once the business grows into a successful organization, it reaches the maturity phase. These organizations are profitable and have a recognizable brand. During this time, operating effectively and efficiently becomes most important. The differentiators that brought on success must be maximized to the fullest to squeeze out profitability. Companies may stay in this "cash cow" phase for many years.

- **Decline:** The decline phase occurs when organizations are no longer increasing in size, profitability, or measures of importance. Decline requires foresight, courage, and determination to identify ways to downsize, divest, or divert efforts to plan for jumping onto the next Growth Curve with a fresh, adapted concept. To handle the leap from the old to the new, the organization needs mature and wise leadership.

THE GROWTH CURVE

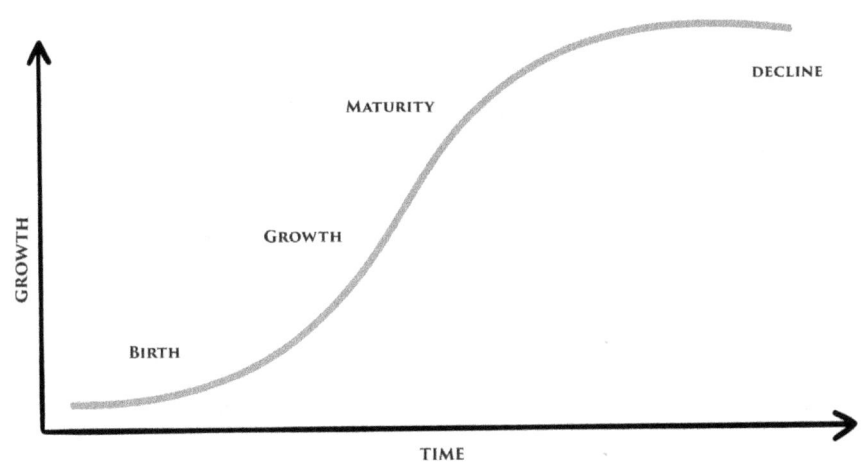

various stages of an S-curve creates different sorts of demands on a leader or on a team of leaders. The four stages along the Growth Curve provide clarity to align a company's current and future needs including leadership, investment, technology, and operations.

My History with the Growth Curve in Business

I first learned about using the Growth Curve for innovation and strategy from Neil Rodgers in 2008. We were involved in advising a renewable energy startup in New York when we learned of each other's affinity for leveraging physics and bioscience to implement change in the workplace. We have been collaborating on business initiatives ever since.

Based in London, Neil has held senior executive positions in the U.K. and worked internationally as an executive coach and organization consultant with a focus on helping executives align their intent and action. He holds several degrees, including mathematics, statistics, strategic management, and professional coaching. Neil's experiences and insights from leading executives through personal and corporate change complemented my own, which emerged from often catalyzing the change or transformation itself.

Neil introduced me to two foundational books: *Always Change a Winning Team* by Peter P. Robertson, M.D., who drew from cybernetics, complexity theory, physics, and ethology, and *Conquering Uncertainty* by Theodore Modis, a renowned physicist. Their work greatly clarified and expanded the theories I already had about change and adaptation as they related to the laws of physics and bioscience throughout all of natural science, such as the Growth Curve life cycle.

I also have enjoyed sharing my findings with and learning from Adrian Bejan, a distinguished professor at Duke University, who has studied this phenomenon for many years and views the S-curve pattern (also known as the sigmoid function) as part of natural flow design. This concept is central to his theory of how all things grow and evolve, whether it relates to energy, rivers, or human populations. His research, conducted with collaborator Sylvie Lorente from the University of Toulouse, France, was published online in the *Journal of Applied Physics*. The National Science Foundation, the U.S. Air Force Office of Scientific Research, and the National Renewable Energy Laboratory supported their research.

As illustrated in the Growth Curve graphic, Bejan and Lorente used the example of a new technology to illustrate their findings. After an initial slow acceptance, the rise of its adoption can be imagined moving fast through established, narrow channels into the marketplace. This is the steep upslope of the "S." As the technology matures, and its penetration slows, any growth, or flow, moves outward from the initial penetration channels in a shorter and slower manner (see the Growth Curve Diagram from prior section).

When I look at nature and at our business cycles, I see the same patterns replicated again and again. Others noticed this before me. "The prevalence of the S-curve phenomena in nature rivals that of the tree-shaped flows, which also unite the animate, inanimate, and human realms," Bejan said. "This theory shows that this is not a coincidence— all are manifestations of the natural constructal tendency of flow systems to generate evolving designs that allow them to flow, spread, and collect more easily." Reading his work reinforced the patterns to which we can see bioscience, and organizations, repeating again and again: tree branches, river deltas, our blood vessels, the branches in our lungs, the growth of human population, and the growth of the economy are all equally predictable. Bejan expanded this work in his 2016 book *The Physics of Life*.[12]

What Bejan proved with natural flow design, Peter Robertson, M.D. bridged into human behavior, with emphasis on business leadership, growth, interconnectivity, and development. There are also natural limits to growth as reflected in the S-curve. Leaders who fail to understand this fact often hang on too long to a mature business and miss opportunities for reinvention.

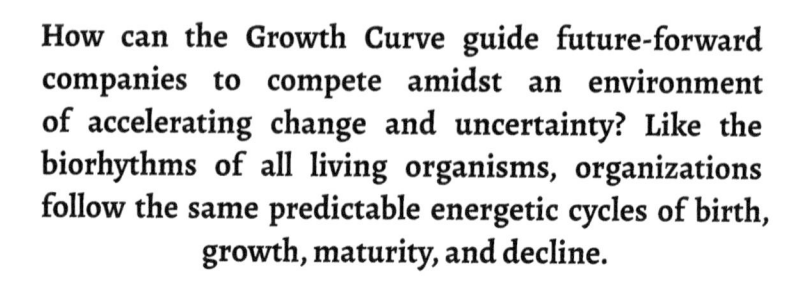

How can the Growth Curve guide future-forward companies to compete amidst an environment of accelerating change and uncertainty? Like the biorhythms of all living organisms, organizations follow the same predictable energetic cycles of birth, growth, maturity, and decline.

The Growth Curve maps each phase, which is replicated during every new strategic direction, product launch, or project sprint. It serves as a bridge between creativity and efficiency, opportunity and risk, startup and stability, as each phase has distinctively different needs for success. Understanding the natural progression through the recurring cycles of birth, growth, maturity, and decline is the key to unlocking your competitive edge and successfully evolving.

Most stakeholders focus primarily on short-term profit generation, but the primary goal of any business should be to fully capitalize on each phase of the Growth Curve, while moving successfully into the next phase. If businesses achieve that goal, they will obtain maximum profitability. In contrast, if organizations rush into the next cycle before leveraging the current one, unrealized benefits will be lost.

Consider the lifecycle of a tree. A seed is planted into the ground, which creates new life that will grow and mature. Once it is well established, the tree is at its most productive phase, yielding its unique combination of fruits, nuts, and seeds for the future. "Success" in nature is the production of seeds that spring up with budding new life of their own over time. At any stage during this process, if enough nourishment is not provided in the proper proportions, the tree will fail to reach maturity. Seedlings are like startup companies or newly released products that compete for resources and growth. Their goal is to reach maturity, then to keep regenerating as much as possible over time, both limited and nourished by their environments.

Companies are living things, no different than plants, animals, or people. The reason nature's Growth Curve is the ultimate predictor and amplifier of success is because it provides clarity about what's needed during each phase of growth to maximize success.

Knowing where a company is located on the S-curve makes it possible to *anticipate* what follows and plan accordingly for the future. With this knowledge, an organization can intervene or invest so that the transition to the next phase will yield greater success. A company's sustainability depends entirely upon how effectively its leadership team navigates from one phase of the Growth Curve to the next.

The following table gives a simplified breakdown of each Growth Curve phase with key strategic considerations. While it is written in the context of organizations, the same concepts may be applied to other lifecycles within a company such as products or services.

Growth Curve Phase	Needs to Maximize Success	Considerations for Strategy
Birth: Startup organizations	Secure funding, form a team that quickly adapts to market signals and recovers from failures, ongoing testing, leaders who naturally anticipate future needs and wants. Take risks.	Identify sustainable growth opportunities, define business model, and generate a clear business plan.
Growth: Organizations with established and tested ideas that are expanding	Establish unique business model, scale operations, establish profitability, leaders who anticipate the future and understand the context of existing market constraints. Take considered risks.	Adapt the business plan to market, establish and clarify brand identity, and begin identifying opportunities for rebirth.
Maturity: Organizations that are profitable and have a recognizable brand	Refine business model, improve operational efficiencies, establish streamlined processes, leaders who can maximize profitability and protect brand. Identify and guard against risks. Actively seek innovation.	Refine rebirth or exit strategies and seek leaders that align with reinvention.
Decline: Organizations that are no longer increasing in size, profitability, or measures of importance	Determine feasibility for continuity, minimize risks, and leaders who can invigorate (innovation, sales, marketing, etc.) or divest (mergers, acquisitions, etc.) Avoid risks to existing business and begin taking risks for rejuvenation.	Activate an exit strategy that will restart the Growth Curve (to continue) or that will divest (eliminate the organization). Take risks for next venture.

Given today's fast-forward pace, the most productive thing an organization can do to achieve success is hit "pause" and identify its current position along the Growth Curve. Traditional "cause and effect" business planning no longer applies, given the current environment of unpredictability. Organizations must consistently redefine what is needed to transition into the next phase of growth.

THE GROWTH CURVE: NATURE'S NAVIGATIONAL TOOL

> **The primary goal of executives today is to recognize and maximize each phase of the Growth Curve, while transitioning successfully into the next phase. If businesses achieve that goal, they will obtain maximum growth and profitability.**

Tracking & Measuring Non-linear, Multi-dimensional Complexity

In reference to the diagrams that follow, S-curves can be drawn mathematically, which is useful when the "growth phenomenon" of interest is trackable using quantitative data. For example, turnover, market share, and profitability, at both product and aggregate levels, are typical growth phenomena that can be modeled using S-curves, tracking either growth rates or absolute values. From these models, estimates of future rates of growth can be calculated. These calculations indicate likely future performance ranges and signal when replacement activities might be needed to supplant the declining phenomenon.

S-curves can also be used in a figurative, non-mathematical way to examine and gain insight into patterns over time in relationships, careers, learning cycles, organization development, phases of projects, stages of innovation, propensity for risk-taking, and many other social, technological, and political phenomena. There is a natural life cycle for growth and decline in all of these systems.

Imagine the complexity of overlaid product, financial, team, market, and other Growth Curves found in an organization. The internal and external factors create an endless, multi-dimensional model with myriad points of intersectionality whereupon risk and opportunity emerge. The following diagram demonstrates five dimensions as an example: capital investment timing, harvest yields, seasons of the year, corporate optimal culture, and optimal profits culture.

POWERING CHANGE

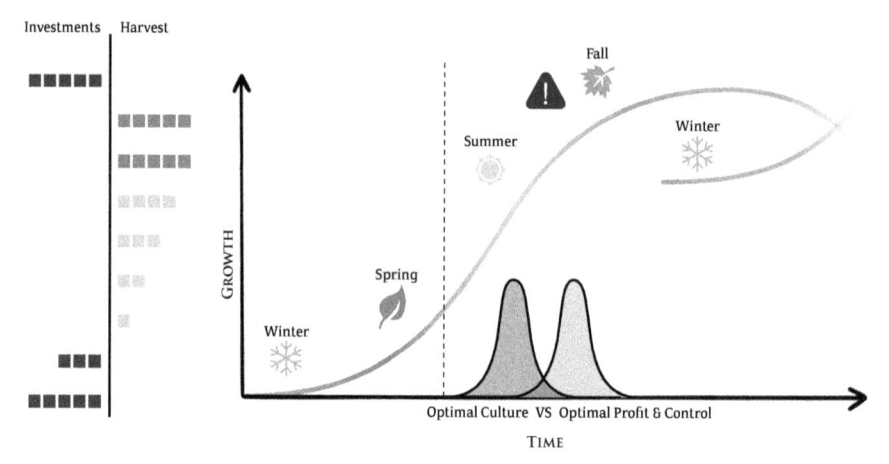

Graphic showing a growth curve with seasons (Winter, Spring, Summer, Fall, Winter) plotted against Time on the x-axis and Growth on the y-axis. Investments are shown declining on the left side while Harvest increases. Two bell curves at the top of the S-curve represent "Optimal Culture VS Optimal Profit & Control."

Source: Graphic from presentation by Peter Robertson, M.D., Human Insight

- Capital investment timing: Investments are largely needed during the birth of an organization, or in the case of farming, during planting. Purchase of seeds, equipment, labor, and tools needed to initiate a new business requires upfront capital as a catalyst.

- Harvest yields: During the growth phase, repayment of the initial investment occurs. Upon reaching maturity, yields of a crop (or a product/business line) are highest.

- Seasons of the year: Winter, spring, summer, and fall occur in the same Growth Curve as businesses. Year after year, they repeat the same pattern, making farming, travel, choice of clothing, and many other seasonally affected decisions predictable.

- Corporate optimal culture: Company culture is strongest when a product or organization is succeeding in the growth phase. It has momentum but has not reached the phase of maturity within which policies, risk-aversion, and specialization into silos has oppressed connection.

- Optimal profits culture: Profits peak later in the maturity cycle, after heavy controls and specialization are put into place to optimize every possible efficiency.

Companies need more than linear business plans to compete today since we are operating in an active, open system. Remember, there are absolutely no straight lines found in nature. Since businesses are living things, why would we expect to measure their growth using a linear, never-ending, one-directional graph?

Giles Hutchins, author of *The Nature of Business*, describes the need for using more natural, fluid ways of viewing business. "Companies of the future are ones that view their organization as a living, vibrant, emergent organism interacting within a living, vibrant, emergent ecosystem." The resilience of an organization is interdependent on the resilience of its business ecosystem. This brings a shift from linear, atomized, supply-chain thinking to interconnected, holistic, ecosystem thinking.

No Growth Curve lasts forever because bioscience works in evolving cycles, and nothing that grows can continue to grow indefinitely. Eventually, and inevitably, decline sets in and it cannot be prevented. In organizations, the question is not how to defend against or how to cushion this decline, but instead how to manage the natural "rhythm of the seasons in a productive way." **A new phase of exploration must begin on a new curve to renew the cycle and continue growth.**

For example, in a sales organization, the natural product life cycle follows a "bell-shaped" curve, where the atrophy curve is a mirror-image reflection of the Growth Curve. To continue growth over a sustainable basis, the objective is to create a new S-curve underneath the apex of the original S-curve by refreshing the original product or by launching a new product, as reflected in the following diagram.

S-Curves in Action

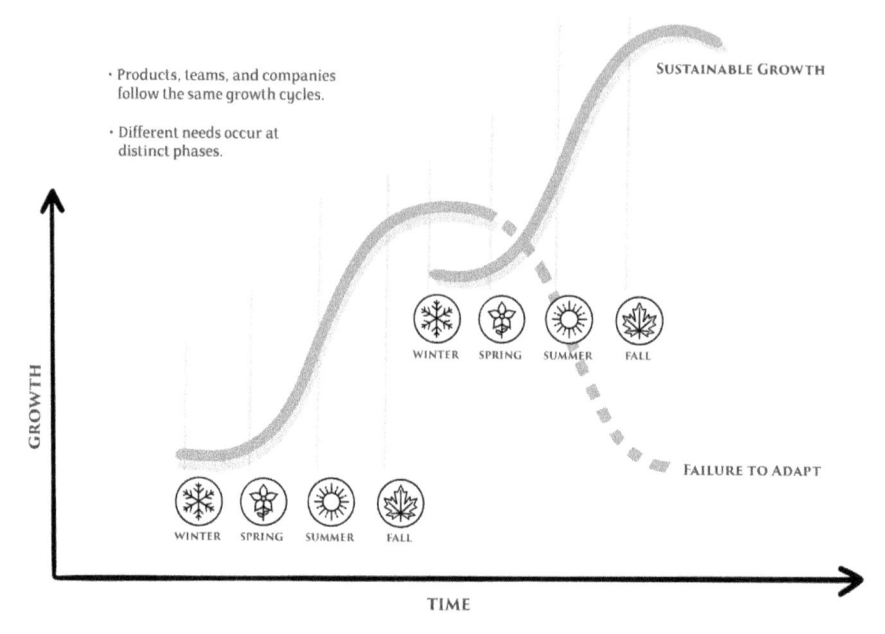

- Products, teams, and companies follow the same growth cycles.
- Different needs occur at distinct phases.

Sustainable Growth

Failure to Adapt

WINTER · SPRING · SUMMER · FALL

GROWTH / TIME

Source: Developed from Human Insight chart.

Key insights may be gleaned from the death of one S-curve and the birth of the next.

- The growth dynamics of a typical product lifecycle follows a "bell-shaped" curve

- If no action is taken, value can be destroyed after it has been created

- A new product is called for, with its own growth dynamics, enabling a revenue stream to be refreshed and maintained.

Leaders must face the trap of staying in the maturity phase for too long. A way out of that trap is to identify opportunities for regeneration well in advance of decline.

Leaping the S: A Roadmap for Regeneration

Regeneration is an important part of the natural world. If we cut our finger, it will heal. If the lawn is cut, it will regrow. In organizational practices, companies are made up of many different series of S-curves; the exact mix will vary according to how you choose to look at the organization's operations. These lifecycles naturally evolve with preference for good ideas, which will follow Constructal Law—meaning they will branch throughout the organization just as rivers form deltas, trees form branches, and our own body forms capillaries. For simplicity's sake, consider a product's evolution over time. Individual product S-curves can form a larger product segment S-curve, assuming the product explores adaptations based upon feedback from customers, markets, and other forces.

For example, taken together in sequence, a series of S-curve product life cycles can trace a larger S-curve, over the full elapsed period, describing the growth and decline pattern of the product class to which the individual products belong. Organizations must constantly define and redefine their approach to reach the next phase of the Growth Curve. Similarly, a product segment S-curve could be one of many product segment S-curves that taken together over time will describe a product sector. And so on. There is fractal-like quality of self-similarity to these S-curves as you zoom out or zoom in and vary the scope of detail under analysis, as reflected in the following diagram (Bejan).

SUCCESSFUL ADAPTATION

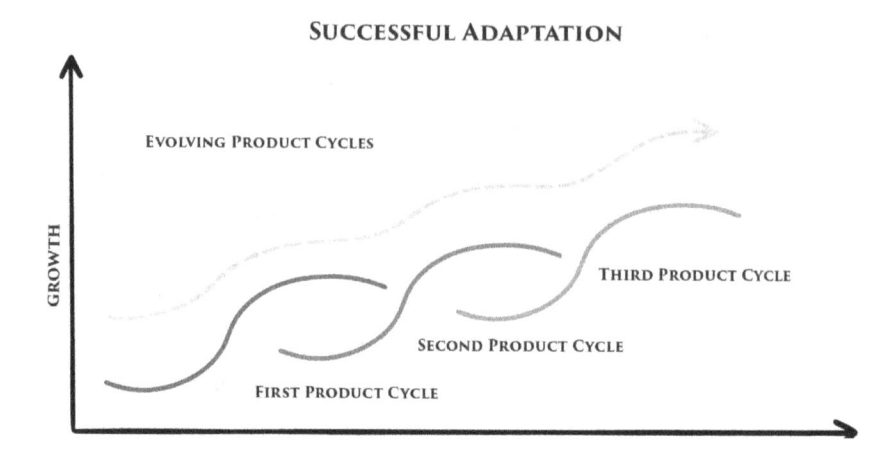

[Graph showing Growth vs. time with overlapping S-curves labeled: EVOLVING PRODUCT CYCLES, FIRST PRODUCT CYCLE, SECOND PRODUCT CYCLE, THIRD PRODUCT CYCLE]

Source: *The Physics of Life*, A. Bejan*
*Note: I varied this diagram slightly to show successive curves beginning between the growth and maturity phases instead of at the beginning of the decline phase to leverage natural momentum and reduce exertion needed for each subsequent startup.

This kind of analysis can be applied to any other quantitative data of interest in the organization; e.g., expenses, earnings, share data, and so on. In complex organizations, with many projects in play, the portfolio of projects can be assessed with S-curves using, for example, Earned Value Analysis as a metric with which to draw the S-curve. Alternatively, using qualitative phases of projects, the portfolio can be analyzed in a figurative way, such as:

- Identify a clear strategy to reach the next phase: the goal is perpetual renewal of jumping onto the next curve
- Assess the attributes needed for the next phase and leadership gaps to grow into it
- Address culture needs and gaps, articulate clear vision with objectives and key results (OKRs), and leverage a simplified dashboard to constantly gauge success and speed

Leaders must pay attention to the portfolio mix of S-curves in play in their organization at any one time. If most S-curves are approaching or are at their peak *at the same time*, then the organization's risks are elevated as its resources will be consumed by these S-curves while contributing little energy to strategic development.

Amazon is a company keenly aware of its product phases and is proactive about continual renewal, which has had a significant impact on its economic health. Founded by Jeff Bezos in 1994 with the goal of becoming "Earth's Biggest Bookstore," it leaned into the new convenience of providing online book purchases, unlike other bookstores at the time that hesitated to transition to internet sales. Amazon quickly reached maturity with book sales, and then leapt onto a new S-curve by selling other products and services. Since then, Amazon has continued to reinvent and diversify itself, perpetually introducing new S-curves with video streaming, cloud computing, and its own delivery system. If Amazon, or any other company, fails to plant new seeds to initiate new Growth Curves, it will fall into decline.

The following chart shows Amazon's various product S-curves as they approach different phases of the life cycle. Note the similar growth pattern over time.

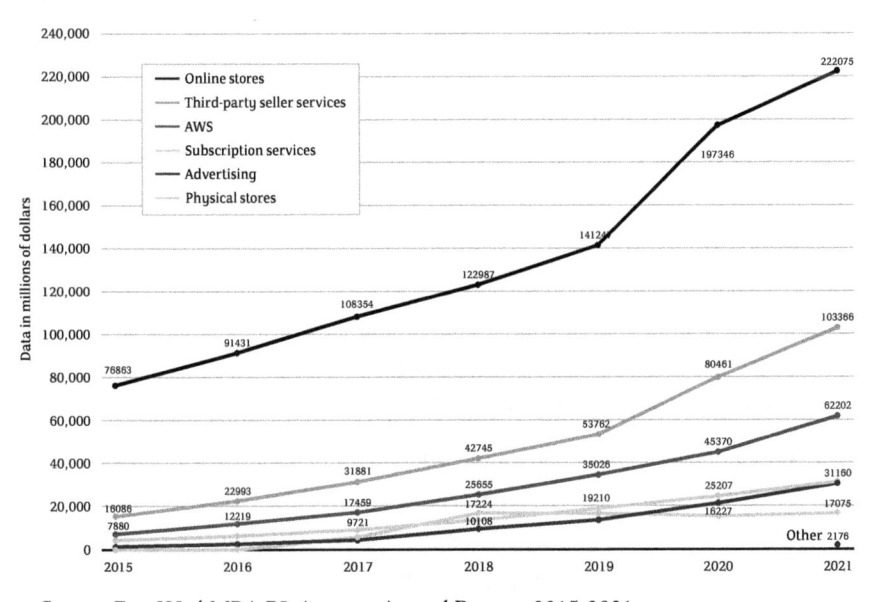

Source: FourWeekMBA BI, Amazon Annual Reports 2015-2021.

To keep that momentum going, smaller iterations of product and service life cycles will ride the backs of each major curve. Imagine each of those S-curves having potentially thousands of sprint (project) level S-curves, each driven by a diverse mix of contributions and interwoven with one another.

When considering the vast building components of a living organization (or organism), we can refer again back to nature and recognize similar structures. For example, consider the shape of DNA, which is arguably the most efficient way to organize sequential data. To vastly oversimplify it, DNA is made of two curved strands that wrap around each other, bound together by data and energy. Can you see the series of continuous S-curves? What's more, the mathematical curve used to predict economic cycles, the oscillating sine wave, is also clearly visible at both the nano scale and the macro scale.

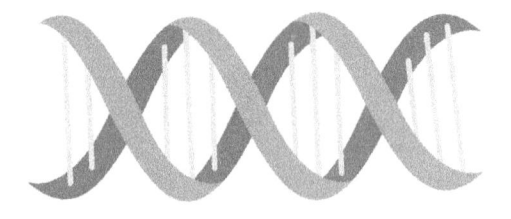

It is absolutely impossible to disconnect a *living* business from the successful patterns in nature, so the further away companies get from the successful processes that thrive in the natural world and have adapted over four billion years, the more disconnected and imbalanced they become.

Remember, modern business has only been evolving for around 300 years. Ignoring these well-established patterns, like running a business by focusing only on the abstract, "non-real" construct of stock value, may lead to puffed-up short-term returns, but will inevitably lead to an organization's death.

These answers from nature cost nothing and are freely available to anyone willing to observe their level of sophistication. Mother Nature is a complete badass with advanced processes that can inform and revolutionize modern business roadmaps far beyond what leaders are taught when earning an MBA. In the same way, **companies already have everything they need to maximize success or reinvent themselves by focusing on what is uniquely needed during the current and upcoming lifecycle phases.**

Consider this concept in relation to two highly successful organizations that used existing resources to disrupt entire industries. According to *Business Insider*, Airbnb is worth more than the top three hotel chains combined. Interestingly, as the world's largest hotelier, Airbnb owns no properties. Similarly, the largest taxi service, Uber, has no fleet of vehicles. Both companies found a way to dynamically leverage assets, manpower, and other resources that already existed. They simply provided the best networks to bring a wide variety of homes, cars, travelers, and drivers into adaptable new business offerings by connecting, directing, and engaging customers.

Change Methodologies over Time

Powering any kind of growth requires energy from a variety of resources. A multitude of frameworks and processes exists to manage change based upon the popular organizational theories of the time. The Growth Curve is a timeless and all-encompassing model that works as a dependable roadmap across all industry types and schools of thought.

The following diagram provides a high-level, visual timeline of widespread, cross-industry adoption of several popular business models that help perpetuate growth. I've prepared it in this way to demonstrate how different models add maximum value in the appropriate phase of the Growth Curve. The limits of growth must be embraced as inevitable fact, recognizing the need to focus innovation on continuous generation of new Growth Curves.

THE GROWTH CURVE AS A BUSINESS MODEL INDICATOR

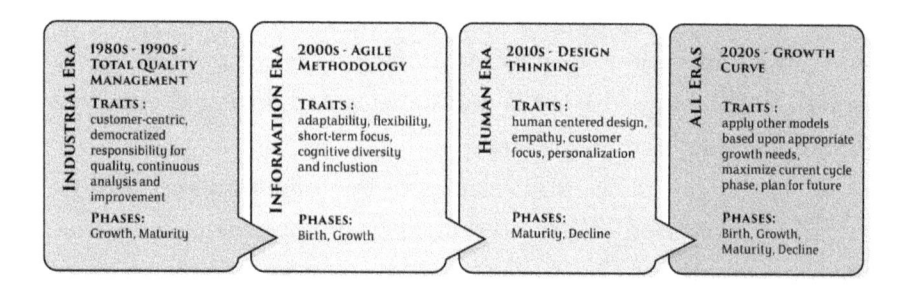

INDUSTRIAL ERA	INFORMATION ERA	HUMAN ERA	ALL ERAS
1980s - 1990s - TOTAL QUALITY MANAGEMENT	**2000s - AGILE METHODOLOGY**	**2010s - DESIGN THINKING**	**2020s - GROWTH CURVE**
TRAITS: customer-centric, democratized responsibility for quality, continuous analysis and improvement	TRAITS: adaptability, flexibility, short-term focus, cognitive diversity and inclusion	TRAITS: human centered design, empathy, customer focus, personalization	TRAITS: apply other models based upon appropriate growth needs, maximize current cycle phase, plan for future
PHASES: Growth, Maturity	PHASES: Birth, Growth	PHASES: Maturity, Decline	PHASES: Birth, Growth, Maturity, Decline

In the 1980s, as the Industrial Era boomed, highly developed countries in North America and Western Europe struggled to compete with Japan's ability to produce high-quality products at competitive costs. During this period of economic disruption, Japan moved to its highest production output since the Industrial Revolution. In reaction, American businesses widely adopted Total Quality Management (TQM) practices from Japan to boost competitiveness. Federal Services in the U.S. also adopted TQM, with key attributes including:

- Quality is defined by customers' requirements.
- Everyone has responsibility for quality improvement.
- Increased quality comes from systematic analysis and improvement of work processes.
- Quality improvement is a continuous effort and conducted throughout the organization.

Because TQM focused on quality and efficiency, it best served organizations or products in the maturity and decline phases of growth.

In the early 2000s, the pace of technological change accelerated quickly in the Information Era. Agile methodology responded by creating shorter life cycles that leveraged functional collaboration among teams. This method was among the first to propose value from cognitive diversity rather than static, inflexible team members. It originally became popular for software development and later expanded to include project management.

Agile life cycles are composed of several short-term sprints or steps to promote speed and adaptability throughout the completion of a project. This allows for ongoing mid-course adjustments and corrections that provide benefits throughout the process instead of just at the end. Flexibility, customer-centricity, and the inclusion of different ways of thinking are tenants of successful agile operations. These methods are most helpful during the birth and growth phases of the S-curve.

Although Human-Centered Design can be traced back to mid-20th century roots, it has become popularized in the Human Era, beginning roughly in the 2010s, with widespread adoption of Stanford University's "Design Thinking" process. Used in ISO standards, Human-Centered Design is an approach to interactive systems development that aims to make systems usable and useful by focusing on the users, their needs, and their requirements, and by applying human factors/ergonomics and usability knowledge and techniques. This approach enhances effectiveness and efficiency; improves human well-being, user satisfaction, accessibility, and sustainability; and counteracts possible adverse effects of use on human health, safety, and performance.

Design Thinking was developed to create new ideas, methods, and products based on the needs and wants of existing users and customers involved. It is therefore most important during the maturity and decline phases of the S-curve to prepare for "rebirth" onto a new Growth Curve.

Today, the Growth Curve transcends prior models as it represents the common lifecycle of all living things, products, services, teams, and organizations. It provides an over-arching guide found throughout the natural world that, when applied to business, can be used to identify which methods will maximize effectiveness during a given moment in time, as well as to plan for everything from human resources to investments and innovation.

Having used the Growth Curve as a guide to develop strategy when consulting since 2017, based upon models created by Robertson in the E.U., I was intrigued to find it as part of the Harvard Innovation & Strategy Executive Course I completed in 2019. The Growth Curve's inclusion in the course curriculum was an important indicator that the concept was becoming more widely understood and adopted by progressive U.S. learning institutions and businesses.

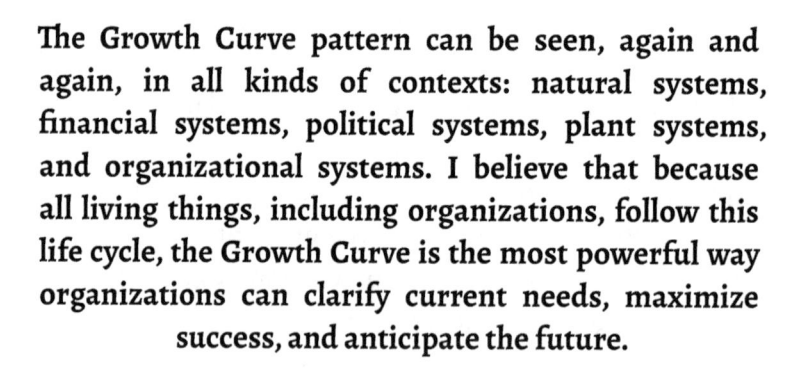

> **The Growth Curve pattern can be seen, again and again, in all kinds of contexts: natural systems, financial systems, political systems, plant systems, and organizational systems. I believe that because all living things, including organizations, follow this life cycle, the Growth Curve is the most powerful way organizations can clarify current needs, maximize success, and anticipate the future.**

Chapter 6

NETWORK POWER: WELCOME TO THE WOOD WIDE WEB

We're all connected; it's like a web. If you touch one part of the web, the vibrations travel all the way to the other end, and you hear a spider swearing.

—GEORGE CARLIN

"Get your head out of your app and look where you're goin'!" I yelled, *as if he could hear me.* Tired, and driving home after work, I rounded the curve where I-40 and I-65 split into opposite directions around Nashville. Out of nowhere, I got a full-bodied dose of adrenaline from Mr. Distracted who veered across the highway in his sporty red Porsche, face buried in his phone and not on the road, as if the other drivers were simply sitting still. Without a glance or a signal, he nearly sideswiped my car as he lurched from the far-left lane to the far-right lane. The driver next to me had to swerve hard to avoid disaster. Honking horns and brake lights lit up the highway in frustration.

I've witnessed firsthand the advantages and disadvantages of Nashville's explosive growth over the last 20 years. Consistently ranked one of the fastest-growing cities in the U.S., studies from Texas A&M rate its traffic congestion the 24th worst in America. Not counting commutes, Nashvillians can expect to lose about 16 hours each year due to increased traffic congestion, according to INRIX's Global Traffic Scorecard.

Highways are a type of integrated network that people and businesses depend upon to thrive every day, and they are but one example of network infrastructure designed to connect and distribute people, information, and things. Networks can be centralized, like highways, or decentralized, like the footpath between your front door and the street. Modern technological networks often have built-in "smart" capabilities, or intelligent decision-making, to enhance effectiveness. They are connected, living systems nourished by electricity and data.

Mr. Distracted might have been using a network-based app like Waze or Google Maps, which disseminated real-time, decentralized information to aid in decision-making. Such an app would have collected live data on his speed, possible routes, and alerts from participating users nearby. Then it would have leveraged predictive analytics to determine current traffic conditions, accidents, congestion, and road construction to suggest the best route. These unseen networks are mostly hidden from view, yet collectively they provide a stream of insights from which decisions may be made.

Societal Infrastructure Mirrors Nature

Whether they're aware of it or not, all drivers integrate seamlessly with multiple external networks. Automated traffic control systems post messages via electronic signage, adjust the streetlights along the roadway, and alter traffic light patterns. Networks display from a vehicle navigation system, a separate device, or a smartphone—each capable of sending more complex information for decision-making like nearby restaurants or fueling stations. Nesting even deeper, those places of business use networks to share their specific hours, contact information, and ratings.

I learned about the importance of properly planning network infrastructure when serving as the chair for the Nashville Urban Land Institute's Civic Leadership Forum, a group that provided independent educational resources about sustainable urban development. Growing cities leverage technology to improve the movement of energy, waste, water, people, traffic, and other things around efficiently. The hardest part, as always, was trying to convince citizens to proactively support the expense of revitalizing their city's infrastructure—most people tend to wait until it is irreparably broken or outdated before they support action.

Networked infrastructure within communities is not new. Perhaps best known for their ancient roads and aqueducts, the Roman Empire used infrastructure to expand into many territories, connecting its armies and supplies with resources. Modern infrastructure provides utility grids, waterways, transportation, and other major networks while also connecting data to decision-makers (both humans and technologies). Beyond basic infrastructure, networks are the fabrics that stitch our businesses together.

The more sophisticated our modern networks become, the more they emulate the billions of years of evolutionary development already found in nature. But what can natural networks teach businesses about rapid adaption, efficient resource usage, and sustainable success?

Ecopsychologist Christyl Rivers, Ph.D commented, "At first thought, the world of business, entrepreneurship, marketing, and more seems a world away from nature. Yet, think about global shipping, complex trade agreements, delivery, and demand. Think about how resources become revenue. All raw materials or innovative ideas must be networked before they become gains for anyone along the supply chain."

Many of the most vital networks developed by modern civilization to support business and industry are buried underground, quietly connecting and providing the things we see above ground to thrive. Communication lines, Internet cables, and electrical distribution conduits as well as water and sewer pipes. These are just a few examples of networks we don't always see; yet they provide critical 24/7 needs.

Think about it: When natural disasters strike, the first data point reported is fatalities, followed quickly by how many people are without power and water. The Department of Energy recently estimated the outages from electrical network failures alone costs the U.S. economy approximately $150 billion annually. For large companies, power outages can escalate into millions of dollars per hour of downtime. According to an article by BloomEnergy, just one day without energy results in the following costs for domestic businesses:

- Data centers: over $12 million
- Car manufacturers: $60 million
- Supermarkets: up to $5 million

Although we try our very best to improve connectivity and effectivity by practicing redundancy, load balancing, demand side management, islanding, diversified generation, and self-healing networks in modern times, we're still no match for the networks found in the natural world wherein:

- Needs are anticipated and actions are taken proactively to prepare.
- Decentralized data is collected in real time, providing important decision-making context.
- Seamless communications yield high connectivity and performance.
- The waste from one form serves as the fuel for another.

Nature's networks have already proven their resiliency in the face of constant unpredictable change. We have much to learn from the behavioral patterns and activity found throughout bioscience.

Fungus Among Us: Lessons from the Mycorrhizal Network

Like our modern networks, mostly hidden from sight underground, fungi exemplify one of the most efficient, smart networks on earth. There are over 1.5 million known fungi species—that's six times more than plants! This incredible species offers many lessons for how businesses can better adapt to change. Fungi are intelligent in that they observe needs proactively, collect distributed data, provide sophisticated communications, and solve problems without generating any waste.

Fungi form an immense network communicating among trees, soil, seedlings, and nutrients by using electrical pulses, similar to our brain's neuropathways. The most common of these networks are mycelia, which contain over 300 miles of web under every footstep (source: *Fabulous Fungi* documentary). Mycelia are incredibly tiny "threads" of the greater fungal organism that wrap around or attach to tree and plant roots. Found nearly everywhere, they are an intelligent species that connects individual plants together to transfer water, nitrogen, carbon, defense compounds, and minerals.

Like the Internet connects multiple networks made up of a vast collection of sites found on the World Wide Web, mycelia collectively compose what's called a "mycorrhizal network." German forester Peter Wohlleben dubbed it the "wood wide web," because it acts in a similar way by connecting trees and plants. Mycelia serve as smart agents that facilitate sophisticated "buy/sell trades" among plants growing under conditions of high resource availability, such as high-light or high-nitrogen environments, with plants located in less favorable conditions, like the shaded canopy floor.[13]

Just as business brokers and those in the transportation sector charge a fee for their services of identifying needs and transporting them, the mycorrhizal network retains part of the nutrients it transfers as payment for its services. This creates a mutually beneficial relationship, allowing all connected parties to thrive. Similar models are used throughout the business world, such as charging merchant exchange fees for the convenience of using credit cards, or ATM fees for receiving cash anytime and anywhere.

Leaders can learn a lot from the mycorrhizal network, which helps forests and the natural world stay connected, just as our Internet, electricity, and other utilities help businesses stay connected.

Lesson #1: Anticipate Needs & Proactively Prepare

Nature exists in a constant state of balance and rebalancing. The mycorrhizal network transfers only what is needed, when it is needed, to maximize the efficiency of the natural system. It anticipates needs, like a lack of light or rain accessibility for certain trees or plants, then transfers the appropriate nutrients to support them within its system. This exchange proficiently handles supply and demand within the underground marketplace of networks.

In recent years, companies have similarly leveraged information from weather reports, traffic impediments, real-time inventory needs, distribution center availability, driver accessibility, and so on to inform supply chain decision-making. These data inputs are considered in various ways when moving goods, some automated through software and others via human decision-making and interaction. Once considered, a reaction or output is generated, and the network sends more data or things into action.

Automation and AI are bringing businesses closer to the types of efficiency found in bioscience. I witnessed this in action when advising a popular lifestyle retailer on innovation and IT modernization. The company used predictive indicators from myriad sources to forewarn of significant weather events, such as a blizzard. We then deployed applications connected to multiple networks to prepare accordingly. For example, a network connected to the current inventory at our retail stores and distribution centers was used to identify and redirect supplies to the local stores where inclement weather was predicted to occur. Shovels, working gloves, salt for deicing, and chainsaws for fallen branches were all added to regular supply runs in anticipation of needs.

Another example common in most industries is the "just in time" inventory practice that anticipates the ebb and flow of supply and demand to improve efficiency. It avoids the need for additional storage of materials not yet needed for production and for products that are not yet needed in the marketplace. It also reduces costs and risk by obtaining, using, and delivering what is needed when it's needed versus creating stockpiles that could lead to unnecessary waste and financial losses.

Lesson #2: Decentralized, Real-Time Data Collection

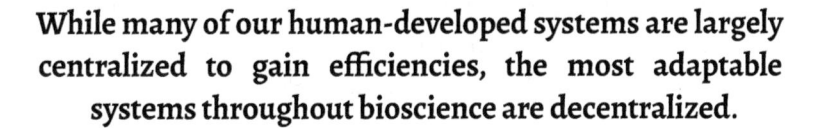

While many of our human-developed systems are largely centralized to gain efficiencies, the most adaptable systems throughout bioscience are decentralized.

In Chapter 1, I discussed how thought leaders like Jobs and Gates helped decentralize computers, making them more widely accessible through smartphones and devices. While mainframes offer more powerful computing, they lack agility and the ability to innovate from a local, user perspective. Due to high-speed Wi-Fi connectivity, information is readily accessible anytime and anywhere in the modern age, especially within the business context.

The value of computing and the World Wide Web (consisting of trillions of pages and portals globally) increased exponentially through activation of the Internet network. It accelerated the ability of businesses everywhere to automate and democratize decision-making locally. Event-driven decision-making that required a manual process of observation and decision-making in the past can now be decentralized through advanced data science.

Businesses juggle a plethora of data from multiple sources, such as purchases, channels, supply chains, web clicks, customer feedback, inventory, and returns. Just as natural networks have evolved over billions of years to automate decision-making and anticipate needs, companies are evolving their networks to get beyond yesterday's computing that required manual interaction. As data is collected from multiple networks ("Big Data"), it can be coupled properly with machine learning (ML) and artificial intelligence (AI) to predict what will happen next. Accuracy and speed are constantly increasing with all types of automation, freeing people to deal with exceptions rather than repetitive tasks.

The transportation sector, as discussed earlier, is full of interconnected, decentralized networks that inform traffic patterns and driver behavior. There are thousands of other examples that apply to businesses, ranging from anti-virus and security alerts to improving operational efficiencies. Just like mycelia in the woods, smart devices collect and report very specific, localized information.

Smart devices that can connect online and exchange data are often referred to as the Internet of Things (IoT). Their decentralized decision-making unleashes opportunities to integrate technology into daily needs. The truncated, brief bits of information transmitted with IoT are similar to a fungi's way of communicating with short bits of data signaling simple yes/no decisions. As human digital communication evolves over time, we are echoing nature's age-old simplicity, favoring texts, "likes" on social media, and "tweets" (now "posts") via Twitter (X) over the long, formal communications of the past.

Local status updates or inputs, as part of multiple complex networks, provide endless possibilities and applications to improve business operations. For example, a sensor in the ground detects whether the soil needs water or not. If it does, the irrigation system kicks on. If not, then the systems remain off. Golf courses, commercial property managers, homeowners, and farmers all benefit from this type of IoT-connected network.

My friend and former colleague Bob Mooney, LEED ID&C, PMP, was under constant environmental and budgetary pressures to conserve and lower expenses during his time as project director for 3,368 family homes at Fort Carson, Colorado. Faced with regional droughts and skyrocketing water costs, he used a variety of waste, water, and energy networks, including IoT, to reduce natural resource use.

When one of Bob's managed communities installed smart technology to reduce water consumption, the benefits were quickly realized. With integrated networks of sensors, the property management team could shut down the system when not needed and save thousands of gallons of water in the process—as opposed to having to drive around the property and manually turn more than 100 timers off and on to accomplish the same result. This initiative both saved manpower and reduced wasteful watering when it wasn't needed. Within one summer, the system paid for itself, and it continues to deliver savings today, over a decade later.

Lesson #3: Seamless Communications for High Connectivity & Performance
A mycorrhizal network looks out for all plants within its network. It practices active "listening" as it intertwines and connects to other networks, creating synergy that helps everything within its reach have a higher chance of sustainable survival. Growing up on a farm, I remember similar automated "sensors" and "decisions" made by various plant species. The sunflowers faced toward the sun to maximize their exposure to daylight; and cornstalks, when parched, would raise

their leaves up toward the sky so that when it rained, they captured the maximum amount of nourishing water.

According to the National Center for Biotechnology Information, listening is critical to performing daily interactions. Its guidance includes the importance of body movement, signals, seeking clarity, and providing feedback—all of which are performed across the wood wide web. As plants have evolved over millions of years, they have become keenly adept at tuning their attention to what is needed or missing.

Businesses, similarly, must actively listen to the needs and wants of their leaders and employees. **A report by Think Talent shows that employees working in organizations with effective communication—ones that manage to minimize the silo effect and centralize communication—are 3.5 times more likely to outperform their peers.** This increase affects day-to-day tasks. As CMSWire reports, 97% of employees believe communication impacts their task efficacy.

It's no secret that businesses improve as their human networks increase. There are myriad ways to interact and connect outside of the workplace: LinkedIn, the Chamber of Commerce, Rotary Clubs, etc. While virtual meetings helped "bridge the gap" during Covid-19, and provide welcome flexibility for many in the workplace, many still feel that face-to-face meetings are necessary. According to a study by the *Harvard Review*, 95% of people say face-to-face meetings are a key factor in successfully building and maintaining long-term business relationships.

Beyond active, personal listening, businesses today are practicing social listening by gathering consumer conversations on social media and utilizing them to uncover insights and inform business decisions. Companies can use social listening to better understand customer needs and wants, which inform its market strategy. According to eMarketer, half of worldwide marketers utilize social listening to understand consumers' changing preferences.

In a similar way, the mycorrhizal network is attuned keenly to potential risks, threats, and opportunities that emerge among all connected parties. In a fascinating culmination of research from academics across the globe, the Harvard Arboretum summarized the ability of the wood wide web to detect potential threats from insects and disease, and then transmit "stress signals" to nearby plants. This informs each species of what's happening on the surface and how it may best change its strategy for the highest probability of survival.[14]

Lesson #4: Waste as Fuel

Above ground, fungi recycle anything that is no longer useful, as nature has no waste. I observed this in a number of ways growing up. Perhaps one of our cats killed, but didn't eat, a mouse. If left alone, fungi would decompose its tiny body, returning important nutrients to the soil. My parents were schoolteachers, so they used day-to-day, random experiences to teach us about science. Mold on bread or cheese was a type of fungi that signaled the food had passed its maturity phase and was in decline (e.g., don't eat it). Eventually, the fungi deployed its enzymes to break down and decompose the food, eliminating waste naturally.

The death of one thing creates the nourishment or birth of another, following the pattern of the universal Growth Curve. *This perfect balance exists because plants, fungi, and animals take what they need to thrive—and rarely more. Excess in nature is reused to create systems efficiency, use less energy, and amplify symbiotic relationships.*

Humans, on the other hand, generate two billion tons of waste every year, according to the World Bank, mostly from businesses, and with an alarming upward trend. Further, the UN Human Settlements Program reports 99% of purchased items are discarded within six months. Waste is expensive; companies pay to obtain resources and materials, then again to generate products. After that, shipping to the consumer adds more expense, and finally the consumer pays to transport used products to the landfill. It's an outdated, linear model still used in non-linear, modern times.

Biomimicry Using the Cradle-to-Cradle Design Model

Imagine a world where we emulate bioscience and turn all that unproductive waste into something financially beneficial. During the 1970s, Swiss architect and industrial analyst Walter Stahel coined the expression "cradle to cradle" while developing a closed-loop approach to production processing. Later, in 2002, German chemist Michael Braungart and American architect William McDonough (recognized as the "father of the circular economy") wrote the book *Cradle to Cradle: Remaking the Way We Make Things*, which introduced their Cradle to Cradle (C2C) design model in detail.

C2C is a design approach to production and consumption based on biomimicry. It considers resources and materials as "nutrients" circulating indefinitely within the economy in a feedback-rich closed loop. It also sees wastes as everlasting resources that could be reintroduced back into the economy. C2C has been an active business strategy for decades that mimics the natural recycling process, so products are "born" with their decline in mind. Yet, most companies have not yet fully grasped or embraced the business value of this concept.

C2C is a fundamental criticism of the popular corporate phrase "cradle to grave," which describes traditional production and consumption models within a linear economy characterized by the life cycle of a product beginning as a raw material extracted from nature and ending as waste materials in landfills. Braungart and McDonough laid down key tenets for integrating the C2C approach within the economy. First is the elimination of the concept of waste. They felt industries should produce materials that could be reused perpetually. Societies should also deploy systems aimed at collecting and recovering the value of these materials following their use. Second is the need to maximize renewable energy. The use of renewable energy sources to include wind power and solar power, including concentrated solar power and photovoltaic technologies, promote the use of perpetual inputs for energy production.[15]

Among the benefits of C2C are economic growth, environmental sustainability, and overall cost reduction. According to research through the Ellen MacArthur Foundation, businesses in the European Union could save up to $630 billion a year by switching to a cradle-to-cradle model and operating through a circular production system.

In the U.S., Shaw industries, a global carpet manufacturer, switched to a cradle-to-cradle business model in 2007. As part of this switch, Shaw achieved a 48% increase in water efficiency and improved energy efficiency, both of which have major environmental and social benefits. Financial benefits of this approach allowed Shaw to save $2.5 million in 2012 alone. Plus, toxic waste and pollution is eliminated through the process, as product materials experience more than one life cycle.[16]

If you want to know what successful business networks will look like in the future, look to the networks that have developed over billions of years in nature. They are adaptive, dynamic, smart, and living. They are efficient, leverage the Growth Curve, and interconnect in a non-linear way based on real-time needs. Natural networks are also resilient, sustainable, and complex—all attributes needed to compete in modern business.

As the complexity of businesses and their dynamic networks increases, organizational leadership must match the level of complexity for it to thrive and maximize growth. Leaders and the systems they manage must co-evolve to adapt within an environment of accelerating change.

Chapter 7

VARIATION: SOLUTIONS FOR RESILIENCE AND SUSTAINABILITY

Diversity: the art of thinking independently together.
—MALCOLM FORBES

One of the fundamental ways businesses can effectively manage complexity and develop solutions for resilience and sustainable success is by emulating the variation found throughout nature, which provides a higher chance of long-term survival. Our greatest teacher is just outside the front door. Variation goes hand-in-hand with adaptability and ethology. Plants and animals that have a high diversity of food sources and living conditions will last longer than those that have highly specialized food sources and temperature requirements.

In the natural world, as in business, variation provides strength.

A Bird's-Eye View on Adaption & Diversity

Consider the adaptability of two types of birds living in the Great Smoky Mountains National Park: black-throated green warblers and cardinals. This class of warbler requires large, undisturbed, and un-fragmented forests to survive and reproduce. In fact, they are rarely found in forests with trees under 60 feet tall. The black-throated green warblers are also indicator species, meaning they are a signal of a healthy forest. The beautiful little songbirds are found nesting and feeding primarily in the upper third of hemlock trees. They have developed this specialization over millennia to coexist with and avoid competition from other bird species.

Unfortunately, the Woolly Adelgid HWA, an aphid-like invasive insect not native to America, is currently wiping out entire regions of hemlock trees across 18 U.S. states. The hillsides in the Smoky Mountains today are covered in barren trees, often referred to as "gray ghost" trees, stripped of life by the insect. Since this type of warblers is dependent on the hemlock trees for its survival, it, too, has begun to disappear. They are at risk of extinction because they have not adapted quickly enough to live in other types of trees.

Cardinals, on the other hand, nest in a variety of trees, shrubs, and thickets. Their diet is widely diverse and can readily be found among forests, suburbs, and city parks. There is even diversity between the male and female cardinals to further support sustainability. The male is bright red, making him noticeable and attractive to the female. In contrast, females are pale brown with only a tinge of reddish color on their wings and tails. This variation contributes to successful mating. Once the baby birds hatch, the female nurtures them from the nest. Her coloring allows her to camouflage herself from predators as she protects the young. These various forms of diversity evolved to make the cardinal a more sustainable, successful species than the black-throated green warbler.

In the business realm, a parallel can be drawn between the threat Blockbuster faced and the threat faced by this class of warblers. At its peak, when the company was in the Maturity phase of the Growth Curve, it employed 84,000 people worldwide and had over 9,000 outlets. Their quantity and variety of titles meant it was king of its industry for a time, but it failed to adapt quickly enough as technology evolved from video cassettes and DVDs to online streaming. Like the songbird, Blockbuster met an untimely fate, entering into the decline phase and missing the opportunity to jump onto the next S-curve. Countless companies meet similar fates by not diversifying themselves to get ahead of the risks—like Kodak, Enron, and Blackberry.

Modern businesses need to be more like the cardinal to better manage complexity, to adapt with mercurial speed, and to evolve with the information age.

DEI & Ontology: The No-Brainer Innovators

In this book, I interchange the terms *variation* and *diversity* as they apply to the *entire ecosystem* of the workplace. **Most organizations limit their diversity initiatives to demographic variations, such as ethnicity, age, gender, disability, and personal preferences.** There is no debate; hundreds of studies already prove that traditional, demographics-based diversity increases a company's chance for long-term success. According to McKinsey, companies that identify as more diverse and inclusive are 35% more likely to outperform competitors. They are also 70% more likely to capture new markets, as reflected in a study found in the *Harvard Business Review*.

Having more variation from which to glean insights improves innovation. From a marketing perspective, a company's decision-makers should have representation that matches its target customer demographics. Otherwise, life needs and customer experience are pure conjecture. Aligning insights leads to an estimated 19% higher revenue

VARIATION: SOLUTIONS FOR RESILIENCE AND SUSTAINABILITY

from more diverse management teams, on average, according to Boston Consulting Group.

This outlook on demographics-based diversity represents a foundational, table-stakes approach that is already protected by laws developed in the 1960s. It is just the *beginning* of how variation in the workplace can expand in modern times to improve business performance.

In more recent decades, inclusion and equity have become important considerations to activate the benefits of strategic variation. What science teaches us about equity is that we are all composed of the exact same energy, so I will not belabor the argument that we are all, in fact, equally valuable. Everyone possesses the same potential to contribute in their own unique way. Failure to recognize this equality hinders an organization's growth potential.

What bioscience teaches us about inclusion is that the context at hand will call for different representation. For example, if a living thing falls into decline, a natural army of fungi, birds, and insects rush to convert the lifeless material back into useful carbon that can nourish new life. If a threatening species approaches, the plant or animal will naturally take measures to protect itself. On the contrary, when opportunities to feed, reproduce, or thrive exist, the natural world will seize them to maximize each phase of growth.

In human civilization, a clown would not be asked to perform brain surgery and a toddler would not be entrusted to construct a skyscraper (unless using Legos)! Specialization within growing businesses was developed to maximize the context needed—accounting, legal services, marketing, technology, and so on. These categories are based upon the outer layer of actions, as mentioned in Chapter 4. I include the graphic concerning energetic balance again here to emphasize how actions, emotions, and energy resources are intertwined.

Progressive organizations are tapping into the power of empathy and the emotional layer of inclusiveness. Many companies use

ENERGY: THE TRUE HUMAN RESOURCE

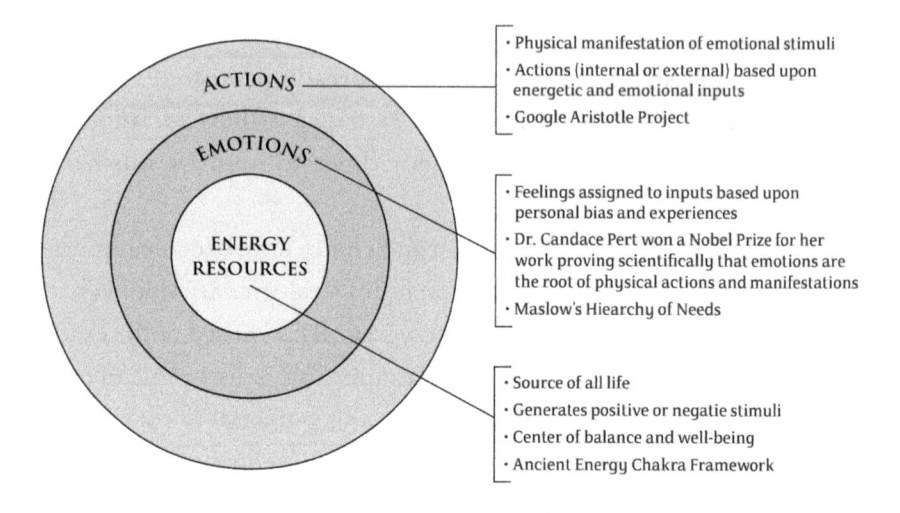

ACTIONS
- Physical manifestation of emotional stimuli
- Actions (internal or external) based upon energetic and emotional inputs
- Google Aristotle Project

EMOTIONS
- Feelings assigned to inputs based upon personal bias and experiences
- Dr. Candace Pert won a Nobel Prize for her work proving scientifically that emotions are the root of physical actions and manifestations
- Maslow's Hiearchy of Needs

ENERGY RESOURCES
- Source of all life
- Generates positive or negatie stimuli
- Center of balance and well-being
- Ancient Energy Chakra Framework

psychometric tools for personal and team development to highlight personal thoughts, preferences, and feelings to better collaborate and communicate. This use of modern psychology has been helpful for companies to understand both internal engagement and enhance communications with external customers.

Beyond psychology, this book highlights the core, energetic resources that drive both emotions and actions. Having served in executive roles that require corporate transformation, idea integration, and habit changes, I have found what has proven to be most effective within those contexts is a focus on ontology. Ontology is the branch of metaphysics dealing with the nature of being. Professional actors become adept at ontology, or being someone else.

One of the first philosophers to consider ontological problems in detail was Plato. Plato made a distinction between the concrete (real) entities that exist in the world and the idealized (abstract) entities that inform them. For example, a sphere is a circular, three-dimensional object. A real sphere will be mathematically imperfect, appearing as a stone, a walnut, or a pea. In contrast, the abstract form of a sphere is perfectly round, like a ball.

In business, we are dealing with mostly abstract concepts such as goals, progress, belonging, and values. Until leaders make those concepts *real* for their teams, they cannot be engaged in the context at hand. For example, a company may list work-life balance on its list of corporate values. The values are eloquently published on the conference room wall and even on little cards given to each employee upon joining the organization.

The values remain abstract, however, and are made real based solely on how authentic the company's role models are *being*. If leadership repeatedly rewards those working overtime and making personal sacrifices in service of the business, they are *ontologically* displaying their value for work-life *imbalance*.

Instead, if leadership rewards those who identify ways to work smarter (not harder), they are making the abstract value of a positive work-life balance real.

When leaders are being ontologically different than the values expected from others, they generate disharmony and imbalance. Such disconnect repels workers at an energetic, core level because *who the leaders are being* does not resonate with the abstract aspirations (values) they claim to value. The imbalance, over time, leads to feelings of disengagement, lack of control, and burnout.

Our balance comes from the natural world, which has been either neglected or overlooked in Western business and values. Humans are unique (in the natural world) in that we allow abstract, non-real concepts to shape our entire emotional and physical condition. It is important, therefore, that we pay careful attention to our choices regarding equity and inclusion.

Ask yourself these questions; they may expand your awareness about natural versus abstract diversity, equity, and inclusion.

> Q: Does my leadership team look, act, or think alike?
>
> Q: Do our reward systems align with our stated values?
>
> Q: What methods are our company using to discover the most valuable potential of each employee, regardless of demographics?
>
> Q: Have I provided clear context about the project, so we know which team members need to be included during this phase of the Growth Curve?

Cognitive Diversity: A Tangible Differentiator

Now, let's shift focus to more evolutionary-related concepts including cognitive thinking, cognitive technologies, and diversity dynamics. For simplicity's sake, when discussing these future-forward concepts, the reader may assume equity and inclusion are a given. They are foundational requirements based on the context of each situation at hand to maximize growth and success.

Beyond demographic diversity, many companies lack cognitive diversity, which is defined as the ways people and systems think. Cognitive diversity rose in popularity in the late 1990s as a response to the overwhelming influx of data in the Information Age, when the accelerating technological pace of change stretched our minds to keep the pace. It expanded "traditional" (demographic) diversity to include the ways *we think* and it considers both humans (cognitive thinking) and machines—bots, AI, automation, and machine learning (cognitive technologies). Without cognitive diversity, companies are limited by "like type" thinking, which stamps out innovative ideas and promotes a "business as usual" environment.

Thinking Diversity

Diversity in thinking provides tangible advantages for organizations. **According to Deloitte's report, "The Diversity and Inclusion Revolution: Eight Powerful Truths," cognitive diversity improved team innovation by 20%.** It means different thoughts, values, and perspectives are taken into consideration to drive faster problem-solving and better decision-making. Studies by Alison Reynolds and David Lewis published in the *Harvard Business Review* in 2017 showed teams solve problems faster when they include diverse thinking types.

Eager to learn more, I became certified to administer the tool used by Reynolds and Lewis to identify cognitive preferences, the AEM-Cube. Peter Robertson, M.D. developed the tool based on the laws of nature, including cybernetics, complexity theory, and ethology. The next chapter is devoted entirely to an in-depth analysis of the AEM-Cube's uses and benefits. I mention it now because it has informed my work and perspective regarding the necessity for cognitive diversity from the moment of discovery.

The AEM-Cube is different than the many psychometric tools I used in the past with leadership development programs, like DISC, StrengthsFinder, Myers-Briggs Indicator Type, PredictiveIndex, and others. These tools effectively address how emotions impact satisfaction and engagement in the workplace, with emphasis on understanding yourself and others for improved collaboration. Most of them are based on Carl Jung's psychology from the early 1900s, and they can be used to improve our sociological well-being in the workplace by tapping into the need to belong and feel loved. It's no coincidence that the father of evolution, Charles Darwin, wrote only twice of "survival of the fittest" in *The Descent of Man*, but wrote 95 *times about love*, and 92 times about moral sensitivity.

Robertson's AEM-Cube was the first tool I had encountered that focuses on what energizes each person, team, and organization ontologically—the physics behind what motivates and moves people into action. Finally, we have a statistically validated way to measure and observe the intra- and inter-personal energy current that powers everything else (emotions and actions).

Sparking Change: Experimenting with Team Dynamics
In 2018, I worked with my colleague, Trent Strobel, to develop a double-blind, live experiment to demonstrate to groups, in real time, the positive results of cognitive-thinking diversity and inclusion in business. We completed six of our own statistically significant studies with billion-dollar organizations during the last few years. Because the studies were performed live, we had a gap during the years when large gatherings were prohibited by the Covid pandemic. In every case, we discovered that teams with varied thinking preferences and members who are energized differently outperform those with like-type thinking preferences in both the number of ideas created (innovation) and the ability to complete an operationalization process (productivity).

Across the six, double-blind studies we performed in billion-dollar organizations, diverse-type thinking teams outperformed like-type thinking teams every time.

During the first half of a workshop, participants were seated with like-type thinkers (based on their placement along the Growth Curve from the pre-administered AEM-Cube tool). It was not revealed how the seats had been assigned. We simply said we wanted to be sure people had an opportunity to intermingle during the workshop. Teams were given two standardized tasks, limited to only three minutes each, that demonstrated innovation and productivity. The time limitation forced participants to leverage their most natural contributing styles.

1. Innovation: Participants were asked to work with their team to identify new uses for alternative ways to repurpose an object (like a pen or paperclip) other than its intended use. They listed all the team's ideas on the front side of a standardized document provided to each table.

2. Productivity: Participants were asked to go through a series of steps representing the phases along the Growth Curve, including:
    a. Which innovative repurpose idea is the best?
    b. Will it work and why?
    c. How do you get into action to make it real?
    d. How will you scale the business and make it more efficient?

During the second half of the workshop, participants were regrouped into cognitively diverse thinking teams. Again, it was not revealed that there was any significance to the seating. The same standard task and time limits were repeated, to keep all variables constant except the use of a different object. The results of our six, double-blind, live studies are in the table below.

Date of Study	Group Type	Size	Innovation Increase	Productivity Increase
June 2018	Fortune 200 Consumer Manufactured Goods Southeast Region Leaders	117	19%	53%
October 2018	Fortune 200 Consumer Manufactured Goods Northeast Region Leaders	48	39%	40%
May 2022	Multibillion dollar global tech company, APAC Region	26	5%	9.5%
June 2022	Multibillion dollar global tech company, Netherlands Region	30	15%	25%
October 2022	Fortune 500 Diversity, Equity, & Inclusion Council	42	12%	43%
December 2022	Fortune 500 US Bank: Marketing Leaders	52	19%	19%

After each experiment, the results were revealed to participants. We led them through a discussion about how working in the like-type team differed from the diverse-type team.

Common feedback from workshop participants was that the like-type thinking group was more in sync and easy to get along with, which is no surprise since leaders tend to surround themselves with people like themselves, and we typically gravitate toward people with similar interests. This tendency is scientifically valid because we're in resonance or on the same wavelength with people like us. In nature, fish and animals also travel in like-type packs to feel more secure and protect themselves.

On the other hand, in diverse-type thinking teams, there is more friction and challenges. This is because different perspectives require conscious inclusion and engagement. In nature, there are many instances of species that develop complementary relationships to help each thrive. For example, birds cling to large grazing animals to feast upon the parasites on the mammal's body, including ticks and blood-sucking flies. This may help keep the mammal's parasite load under control, while the birds get an easy meal. The remora fish and sharks are another example of how different contributions benefit both. The fish obtains nourishment by eating parasites from the shark's belly. In turn, it receives scraps from the shark's prey and protection against other species.

As we continue this research about cognitive diversity, we are learning and demonstrating what positive impacts strategic variation generates, in both innovation and productivity.

Cognitive Technologies

As a complement to diversity of thinking, the integration of automation, machine learning, robotics, and artificial intelligence is transforming efficiency, accuracy, and productivity. Automation is pouring into the workplace at an accelerating rate. Cognitive technologies involve robotic process automation, traditional machine learning, natural language processing, and rule-based expert systems. We experience artificial intelligence (AI) in our everyday lives as it handles answering services, account balance inquiries, and payment services, plus ChatGPT has revolutionized accessibility for millions. Those dreaded voice prompt systems to direct customer calls, the chat bots on retail websites, the sensors signaling to stores that you've arrived in the pickup spot, the automation of payable processes, and countless other applications are incorporating machines into our teams and workforce.

One example is NASA employee George Washington, who handles clerical work more quickly and efficiently than anyone in his position has before. He never complains about working late hours or needs a cup of coffee. George is a bot who takes in information from emails and identifies job candidate suitability—something formerly completed by HR personnel. He represents a new type of cognitive diversity through the implementation of technology.

Most companies begin by integrating AI and other machine systems to replace the repetitive, "low-hanging fruit" projects. This allows them to achieve quick wins early in the adoption process. Accepting these new "employees" at work as equals, however, can initially cause concerns about long-term job security.

According to research from Accenture, 86% of marketers believe AI will make their industry's work more efficient and effective. Practical applications already include processes like precision targeting, dynamic ad creation, and marketing automation. In these cases, adopting cognitive diversity into the workforce increases accuracy and frees humans to focus on more complex tasks.

Data from Emarsys shows widespread inclusion of AI by retail marketers worldwide:

- 54% Personalizing customer experience and behavior across channels
- 52% Managing real-time customer interactions
- 48% Identifying or reorganizing customers across channels
- 41% Targeting appropriate audiences for new customer acquisition

In NASA's case, cognitive technologies, through the inclusion of bots, is one way to meet shrinking budgets and allow existing humans to focus on higher-level tasks. Thus far, George Washington has been met with open arms, and NASA plans to bring on a Thomas Jefferson bot soon. Their experience shows that integrating machines as cognitive tools has improved the bottom line.

Embracing cybernetics, the science of communications, and automatic control systems in both machines and living things may sound like science fiction, but in today's business climate it can significantly boost agility, innovation, and profits. As cybernetic models improve, they will become more commonplace when organizations consider the functions of diversity and inclusion executives.

Diversity Dynamics

> While cognitive thinking and technological diversity can identify where individuals and machines contribute most powerfully to boost engagement, productivity, and innovation, Diversity Dynamics puts it all into action. Diversity Dynamics is the active, ongoing practice of aligning team members most powerfully with the context at hand.

Tech teams may relate to this practice with project sprints, where workers are often shuffled based upon the rapidly changing needs of each development. However, to compete with giants like Amazon, *every* department within an organization must now be fluid, agile, and responsive.

In 2018, my colleagues and I began practicing Diversity Dynamics to help companies define how to apply the principles of variation, as found in nature, to improve engagement and growth. This expanded beyond cognitive diversity to also include how we each get into action ontologically, based on what intrinsically drives each team member forward. It is a dynamic practice of leadership that changes based upon the phase of the Growth Curve and on the particular business segment's purpose (e.g., marketing teams will be skewed in one direction while engineering teams will be skewed in another direction).

> **Everyone is unique and energized differently. To maximize growth and engagement, companies must identify and align those contributions where they serve most powerfully. In doing so, they can accelerate innovation, agility, and productivity.**

The key steps to leverage diversity dynamics are to:

1. Assess: Identify how individuals are naturally energized. For simplicity's sake, we will focus on preferences within the business contexts of adaptation to change, complexity, attachment, and purpose.

2. Align: Evaluate the mix of preferences on each team to ensure alignment with the context at hand. This could be a specialized function within the organization, a project, a service, or any number of purposes for which the team is responsible to execute.

3. Adapt: Adjust the mix of preferences on the executive leadership team to ensure a balance, emphasizing specific needs from each phase of the Growth Curve at hand.

The techniques described here are based on the evidence that variation in nature and in business equates to value; it increases the likelihood of long-term survival. Further, the truths we need to guide businesses forward more effectively are right in front of us in nature; we just need to pause long enough to observe and emulate them. The remainder of this book will focus on unleashing that success through the use of Human Dynamics to discover hidden potential, boost productivity, and increase profits.

Part III

UNLEASHING POTENTIAL ENERGY

Chapter 8

THE AEM-CUBE: THREE DIMENSIONS FOR CHANGE AND GROWTH

*Always remember that you are absolutely unique.
Just like everyone else.*

—MARGARET MEAD

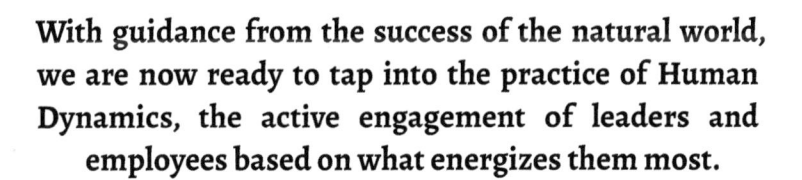

With guidance from the success of the natural world, we are now ready to tap into the practice of Human Dynamics, the active engagement of leaders and employees based on what energizes them most.

Kate Nicholas was the gregarious VP of Sales for a Fortune 500 consumer product manufacturer. The company made many products that were sold to modern and traditional retailers alike, with cigarette lighters being one of their best-sellers for over 60 years. One such customer was Family General, a sprawling retailer with traditional markets in rural communities across the U.S. Their margins were paper thin, so their adoption of new technology lagged behind most of the retail sector by at least a few decades.

Family General's demand for lighters was strong but not easily predictable according to traditional methods. This meant that Kate's team had a recurring problem. They couldn't reliably predict when to resupply lighters and how many to re-stock, creating "out of stock" situations on a frequent basis. As a result, customers looking for lighters at Family General would sometimes need to take their business elsewhere. In more modern retail outlets, software automatically alerted Kate's team when reorders were needed, but older stores like Family General still often relied on managers to do inventory manually, when time permitted. As a result, the lighter company lost over $2.5 million in sales from "out-of-stocks" in Family General stores during 2016.

Kate was determined to find an innovative solution to prevent out-of-stocks and recapture millions in lost revenue. She thought she knew just who to call, a team out of the Northwest known for young, tech-savvy reps who consistently bested other regions when it came to innovation and creativity, working mainly in urban and suburban areas. And, like all client teams, they were incentivized by increased product sales. Finding a solution would mean more money in their pockets, so they were eager to solve the problem.

Over the months that followed, the Northwest team tried a number of new solutions—none of which panned out. First, they proposed rolling out new inventory software to Family General that would automatically alert its managers that supplies were getting low and prompt re-orders. But, this was a no-go for the rural markets that preferred low costs to new technology investments.

Next, the team proposed providing Wi-Fi-enabled mobile devices for each store to scan and track inventory more quickly and consistently. On the surface, it seemed like this idea had merit, and a pilot was launched with 100 stores.

Unfortunately, the Northwest team had not accounted for the fact that many of these stores were in regions so rural that internet connectivity was not readily available. Further, Wi-Fi coverage within the store often had "dead zones," rendering the technology useless. What's more, there was a training barrier—Family General employees felt like the scanning device was just one more chore in an already overly busy day.

Now well into 2018, a very frustrated Kate reassigned the problem to her company's Midsouth client team. This team had longer-term relationships with their retail clients and were known as being more cautious and risk-averse. Understanding their stakeholders, the Midsouth team knew that implementing new technology was unlikely, so they examined what they already knew about how Family General's people, technology, and processes interacted, looking for ways to streamline reordering and increase accuracy without requiring new investments.

After interviewing many managers and visiting the stores to evaluate existing processes, the Midsouth team quickly realized that the solution had been there all along. Shrinkage, or theft, was accounting for most of the problem, so the team suggested moving the lighters behind the checkout counter alongside cigarettes and other items that tended to "walk away." This simple shift reduced theft and gave managers higher visibility when supplies were running low.

What the Northwest team failed to resolve after 18 months of introducing new methods, the Midsouth team was able to resolve within a few weeks. Kate's first thought had been to go to the team who favored innovation via brand-new solutions. But, in this scenario, the right team for the job innovated via refining and improving an existing process. Both contribution types are equally valuable but every problem is different; some need to be solved with creativity, others efficiency. Some require new technology; others benefit from empathy. As a leader, Kate's job was to assess the issue and then align the right mix of people to find the best solution for this specific problem and context. If there's no connection between a team's strengths and the problem at hand, one might as well be throwing darts at a board, hoping one of them sticks in the right place.

The Connection Point: Alignment at the Core

Behavior researcher and business consultant Dr. Brené Brown defines *connection* as "the energy that exists between people when they feel seen, heard, and valued; when they can give and receive without judgment; and when they derive sustenance and strength from the relationship." When a connection has been made between the source of power (the employee) and the beneficiary (the organization), the potential exists to amplify or add momentum to work in progress. When connection does not occur, people disengage, burn out, and quit. Think of it in energetic terms. You may have a drawer full of batteries in different shapes and sizes, each with a unique purpose. Unless the power source is placed inside a device that aligns perfectly, it sits idle—and eventually loses power.

People are no different. It takes many unique and diverse resources to power a multitude of devices. **Creating engagement starts with being connected to a shared purpose. Every employee must be able to identify how their actions directly relate to the desired outcomes.** If they don't feel like their contribution is necessary or valued, their power is minimized and ultimately depleted.

According to the U.S. Energy Information Administration (EIA), up to 60% of energy used for electricity generation is lost in conversion. What's more, an additional 5% is lost during transmission. For organizations, "line loss" is measured by inability to adapt, imbalance, burnout, and ultimately turnover.

With hybrid work, 24/7 emails and news cycles, and the accelerating pace of change, it may not come as a surprise that Career Builder reports 75% of employers report losing two hours of productivity daily due to employees getting distracted. But distraction due to an overload of competing information is not the only issue; distraction also occurs when objectives and expected results are unclear or ambiguous.

In Jim Collins' legacy book, *Good to Great*, organizations with returns of three times the market over a 15-year period had effective leaders who catalyzed commitment to a clear and compelling vision, which stimulated higher performance standards. But it goes beyond aspirations. As Malcolm Gladwell points out in *The Tipping Point*, we need both relevance (personal connectivity to the cause) and less than three clear, actionable steps.

By leveraging Human Dynamics, we have found the highest impact way to maximize engagement and productivity: set a clear direction, then work with each individual to identify one clear action to move forward at a time. One action with full-powered focus has less "line loss" than multiple new requirements at the same time, thus diminishing the same energy. Further, try to integrate that one action into a process that already exists, such as an annual review or monthly touch-base agenda, because it creates less friction and resistance than establishing a completely separate deliverable. Plus, it will be easier to track progress year over year.

Purpose can also boost productivity, yet a recent study by Price Waterhouse Coopers reported that only 28% of employees surveyed felt connected to their company's purpose and only 39% said they could clearly see the value they personally contribute. That leaves a lot of productivity on the table, since Gartner reports the majority of employees cite purpose among the highest reasons to engage at work. While purpose has always been a primary need, the ripple effects of the pandemic swelled its significance.

In Human Dynamics, once a person is connected and clearly understands their direction, purpose becomes the catalyst that gets them moving. The higher their vibe, meaning the more they are naturally aligned with and energized by the role they have been assigned, the more potential energy is released. This creates more engagement and higher productivity.

> The three steps of connection, direction, and power through purpose are critical to unlock greater business agility, engagement, and innovation. Putting the principles of Human Dynamics into action starts by discovering how each person is most energized to connect with their work, their team, their organization, or themselves. It is the first step to engaging the human *resource*, which is their (largely untapped) energetic power.

How each person is most energized to connect can be measured by the AEM-Cube. Briefly introduced in the previous chapter, the AEM-Cube was developed by Peter Robertson, M.D. and was based on 20 years of research into the scientific components and dynamics of effective teams and individuals. His research included over 100,000 participants worldwide to provide scientific validation across a wide range of industries and cultures. Using attributes from cybernetics, complexity theory, ethology, evolutionary theory, and physics, the AEM-Cube draws from the laws and balance found in nature.[17] Since we, like every living thing, are made of energy, the laws of bioscience apply to our actions and reactions.

Most modern businesses already understand and account for the psychological feelings experienced by employees, but my work goes a layer deeper into the core that generates emotions and actions. Maximizing energy potential at work is driven by ontological sociobiology, how we are all energized in different ways that are equally powerful. Individuals feel connected and thrive most when they are doing what aligns most with their inner core. Further, when teams align complementary ways of engaging, they maximize success for the task at hand—leading to greater productivity, innovation, and profits.

In addition, working with personal purpose or "flow" allows workers to tap into the energy current that runs within them to avoid burnout, increase engagement, and improve retention.

How It Works

The first step of unleashing this potential energy is to find it, but how can we accurately measure each person's unique preferences for connecting in an empirical, quantitative way? The AEM-Cube identifies an individual or system's peak potential engagement point by measuring their internal stressors and motivations. Different than traditional psychometric tools, it is designed for use in business environments of transformation and rapid change where adaptability and alignment are critical to compete.

Traditional psychometric tools work well to predict *emotions* or thinking. **I developed Human Dynamics, however, to address the ongoing, active management of personal, team, and organizational energy, which is the catalyst powering its growth.** Based on two decades of successfully introducing and implementing change and having deployed each of the most common psychometric tools at one time or another, we found that the AEM-Cube tool was more effective at consistently predicting how people and teams will engage, manage complexity, and react to change.

The AEM-Cube presents a visual tool that showcases three crucial sociobiological measures. The first two metrics draw insights from ethology—a scientific study focused on behavior in natural settings. This perspective sees behavior as something that has evolved to help us adapt over time. A pioneer in this field, John Bowlby, highlighted how ethology outshines traditional psychology in predicting human behavior. This means leaders can use an understanding of ethology to choose the right people for specific tasks. In simpler terms, it's the art of selecting team members who are naturally well-suited and energized by certain types of jobs.

Peter Robertson, M.D. defines ethology as "the science that, using the perspective of evolution, investigates how and why a specific behavior develops and what drives that behavior." It's like peering into the history and driving forces behind why we act the way we do.

The AEM-Cube examines three key dimensions of adaptation—Attachment, Exploration, and Managing Complexity—to identify insights into the natural and unique contributions each person brings to the process of change. The key values of AEM-Cube results are to identify and deploy untapped potential by aligning how someone is naturally energized with the role they play in a job. This alignment translates directly into higher engagement, innovation, and productivity. When this alignment is off, people feel drained because they are being asked to perform tasks that take greater energy to stay in a balanced state. The longer someone is asked to work in a role that does not align with their preferences, the higher their risk of burnout.

Each science-based dimension—Attachment, Exploration, and Managing Complexity—is mapped in a 3-D model with an axis containing values from 1 to 100 to represent the percentile scores of each participant compared against the norm group. Every place along each of the three axes is equally important and has the potential to contribute powerfully within the right context of an organization. Working in a role that is opposite to how you are naturally energized for long periods of time requires you to exert greater energy because it does not align with your natural vibe. This is a visual representation that can reliably predict how burnout happens.

The following table shows my personal results relative to the norm.

Dimension	Percentile	N = 1
Attachment	61	
Exploration	98	
Managing Complexity	87	

- Attachment—My score falls at the 61st percentile, meaning I prefer roles with human interaction to those focused only on content. Since it is within the middle third of all scores, burnout is less likely for me when working with processes, things, or content than someone with a more extreme score.

- Exploration—My score for exploration, or preference for change, is quite extreme at the 98th percentile. This reinforces my attraction to roles where I am responsible for igniting innovation and transformation. Recognizing this extreme preference opened my eyes with appreciation for the vast majority of others who are not as excited by change as me, and it made me cognizant of the need to consult someone on the opposite end of the spectrum when making important decisions to reduce risks.

- Managing Complexity—At the 87th percentile, this score was also considerably skewed toward generalization versus specialization, which explains why I thrive most in roles where one must understand a moderate amount about a lot of different things. When making enterprise-wide decisions as CEO or chief innovation officer, it required a strong ability to see how changing one thing would impact everything else. Over the years, I consciously recognized this tendency as being most interested in the "big picture," so it was a key reason why I pushed myself to endure more rigorous, specialized education and certifications to help reduce the risk of making decisions without fully understanding their implications.

Placing these three axes together, the tool visually forms a cube shape with a mathematical coordinate that describes each person's peak potential for performance. This cube represents my coordinates as a basis to help understand the description that follows.

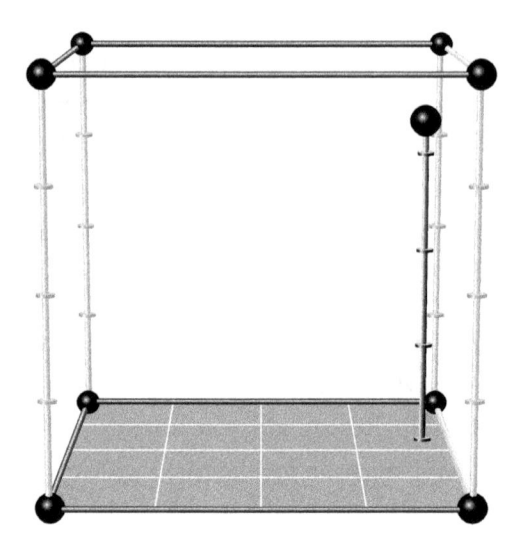

The Attachment Axis goes from front to back, the Exploration Axis goes from left to right, and the Managing Complexity Axis goes from the floor to the ceiling of the cube.

Attachment Axis: Relationship—Content Focus

One charming example of attachment is the story of ducklings and their mothers. Ducklings, soon after hatching, imprint on the first moving object they encounter, which is usually their mother. This profound attachment plays a crucial role in their ability to thrive. The mother duck serves as a source of protection, warmth, and guidance, leading her ducklings to suitable feeding areas and providing a secure environment. Attachment ensures that the ducklings stay close to their mother, benefiting from her experience and learning essential skills for finding food and evading predators. The bond formed between mother and ducklings boosts their chances of survival and helps them develop the necessary social and navigational skills for a successful life. It's a remarkable illustration of how attachment in ducklings is intricately linked to their well-being and growth in the natural world.

In humans, attachment is the instinctive drive to develop bonds with either "people" or "matter" to derive a sense of security. Attachment orientation develops during the first two or so years of life, when individuals "attach" to someone or something that behaves consistently in their world and gives them a sense of safety. Successfully developed, attachment promotes the conditions that enable individuals to explore, learn, and adapt to their daily life environment.

People attachment is related to using human relationships as the primary foundation for security, while matter attachment is related to using non-people-related areas of focus to derive security. The word *matter* includes in its scope a wide range of options, from tangible objects, like computers, to non-tangible concepts, like scientific theories or processes. For example, as a small child, a favorite pet or toy may bring the most comfort to matter-attached people, while those who are people attached would glean security from being with a friend or family member.

Understanding a person's attachment preference is important because it guides the approach most likely to connect with and engage them. Both types are equally critical for sustainable business success. **Identifying and focusing each type of attachment on projects that align well with their natural contribution preferences unleashes greater potential for personal and team success.**

People or Relationship Attached

- Peak Contributions: Developing relationships, collaborating with others, and empathizing.

- Personal Connection: Will engage most effectively with others, especially face-to-face. If isolated from others for an extended period, they will feel drained. If isolated for too long, for example during the Covid-19 lockdown period, those workers who are strongly people attached will experience higher rates of burnout.

- Activate and Direct Potential: Make time to meet face-to-face when discussing important topics. Turn on your video when meeting virtually. Calls or conversations are more likely to yield action than emails or written notices. Work well in traditional or hybrid environments that allow collaboration with others. Provide options for remote employees to work in a shared office space or somewhere they may receive at least some personal interaction. May also prefer in-person learning.

- Common Business Roles: Jobs that involve an understanding of and/or interaction with people, such as customer experience, public relations, customer service, employee experience, team building, human resources, customer behavior, and client relationships.

Matter or Content Attached

- Peak Contributions: Developing or working with processes, systems, or things. Research and other types of work with products, things, or services.

- Personal Connection: Will engage most effectively when working with the things they love. Communicate in writing when discussing important things or decisions. Be sensitive to requiring video use when meeting virtually; accept they may not want to be distracted by it.

- Activate and Direct Potential: Work well in an environment where they may focus on the things they find interesting. May thrive in remote settings, if the proper focus and tools are available, but may also thrive in an office setting if allowed to spend time developing the processes or things that energize them. Demonstrating an appreciation for their specialty, for example a product, service, process, or application, will yield greater connection and engagement. If they participate in team or company gatherings, they will be more energized by having clear things to do or discuss versus just getting together to chat. May prefer online or self-guided learning.

- Common Business Roles: Jobs that involve an understanding of and/or interaction with matter, content, or things, such as research, systems development, engineering, process improvement, risk analysis, programming, maintenance, product design, and data analysis.

Exploration Axis: Exploration—Optimization

In the animal kingdom, the contrasting risk-taking behaviors of cheetahs and zebras provide a captivating example. Cheetahs, as supreme sprinters, engage in high-risk hunting strategies. They rely on their incredible speed to chase down prey, which demands an immense expenditure of energy and increases the chances of injury. This bold approach can lead to spectacular successes but also to failures. On the other hoof, zebras, the favored prey of cheetahs, exhibit risk-averse behavior. They gather in large herds and utilize their black-and-white striped coats to create visual confusion, making it difficult for predators to single out an individual target. This safety-in-numbers strategy helps provide security, as the odds of any individual zebra falling victim to a cheetah are considerably reduced. This divergence in risk-taking strategies between the apex predator and its prey showcases the intricate interplay of how different species evolve behaviors to best suit their survival needs and ecological niches.

Exploration is the instinctive drive "to go beyond what we currently know, without necessarily knowing what we will find." The instinct to explore is innate in human beings. The ability to explore and tackle new frontiers, to go beyond what is currently known, is essential for survival. The instinctive conviction of exploration is that there might be profit in any unknown situation without the need to know what the profit will be. The Exploration Axis corresponds directly to the S-curve, serving as a helpful indicator for teams by balancing those who are risk takers with those who are more risk averse.

The following graphic reflects my results, when mapped by the AEM-Cube along the Growth Curve.

Exploration   Optimization

People vary as to the levels of exploratory behavior they exhibit. Those with a highly exploratory nature, like me, will be drawn toward the unknown and areas of potential interest, constantly seeking new ways of doing business or new people and things to discover. They experience change as an abundance of opportunity to experiment.

Individuals with more of an optimization orientation will tend to value past experiences when faced with new situations or dilemmas. Prior learning of what worked successfully in the past, or proven frames of reference, are used as the basis from which to interpret and make sense of new encounters. Their disposition when faced with disruption or change is to reinforce what they currently know, bolstering their ability to cope with the new to provide stability.

Some team members are natural catalysts, like cheetahs that thrive on the unknown. Others are energized by identifying risks, like zebras that prefer stability to chaos. The key is to identify each person's peak contribution point, or the place along the Growth Curve where they feel engaged, included, and happy. Then, make sure they are placed in a role that aligns with and leverages their greatest potential.[18]

Based upon the phase of life cycle, overall corporate leadership must be the right balance of change preferences—including both explorers and optimizers. Various teams within the company, however, may need to be skewed in one direction or another. For example, most engineers tend to perform best in roles where they will practice optimization, efficiency, and finding risks. Purdue University professors Scott Hutcheson and Ed Morrison found the vast majority of the school's successful engineering students scored on the optimization end of the Exploration Axis (the top part of the Growth Curve). On the other hand, studies from Human Insight show those responsible for research and development tend to be exploration focused, with preference for exploration, creativity, and change. This is because they are naturally energized by change and the unknown possibilities.

Explorers

- Peak Contributions: Adapting to change, exploring new solutions, generating ideas, and taking risks.

- Personal Connection: Will engage most effectively when given the freedom to create and innovate.

- Activate and Direct Potential: Work well in an environment where they have support in taking risks without too much formalized process. Energized by projects in the inception or birth and growth phases of the Growth Curve. Allow opportunities to change, create, and disrupt the status quo. Will burn out if placed in a risk-averse environment without the ability to contribute to change for too long.

- Common Business Roles: Jobs that require original or innovative ideas that have not been tried before. Also, roles that require constant agility, like strategy, research and development, sales, entrepreneurs, innovation, architecture, and design.

Optimizers

- Peak Contributions: Protecting the brand, scaling, structure, stability, efficiency, standardization, and security.

- Personal Connection: Will engage most effectively when trusted to identify and mitigate risks. Provide clear expectations and position change through the lens of avoiding risk instead of a new opportunity.

- Activate and Direct Potential: Work well in an environment where they can refine, scale, and improve existing products or services. Contribute most powerfully to the later stages of the Growth Curve, maturity and decline. Provide a stable work environment with clear and established processes where employees are rewarded for finding risks, efficiencies, and improvements. Will burn out if placed in an environment of constant change without established processes for too long.

- ♦ Common Business Roles: Jobs that require efficiency, standardization, and compliance, like engineering, safety, security, legal compliance, human resources, programming, and operations.

Managing Complexity Axis: Generalization—Specialization

Complexity characterizes the behavior of a system whose components interact in multiple ways and follows local rules, leading to nonlinearity, randomness, collective dynamics, hierarchy, and emergence. The term is generally used to describe something with many parts where those parts interact to create synergies.[19] The complex systems found in businesses need leadership that mirrors their complexity to thrive.

Managing complexity is defined as the way people apply their life experience to cope with ever-increasing amounts of complexity in their environment. It is more developmental than instinctive in that we can increase our ability to cope with complexity the more of it we choose to experience.

People scoring on the lower end of the managing complexity scale tend to approach the world from their individual perspective, through the specialized skills and competencies they have developed, and how they can bring these to bear in the environment they occupy. Whether challenges, problems, and solutions are simple or complex, they will tend to approach these from the perspective of their individual competencies and skills.

Associated with a high level of individuality, specialists will often exhibit high energy and creativity, expressing their competencies with a strong focus on personal knowledge and/or skills development, competitiveness, and even perfectionism in their endeavor to be the best they can be.

People scoring higher on the managing complexity scale tend to approach the world from a generalized, team, organization, or systems perspective. They tend to exhibit higher perception of the entire environmental context, openness to outside ideas, integration, overview, and focus on stimulating teamwork. Their attention is likely to be less on their personal competencies and more on their integrative contribution.

Generalizers

- Peak Contributions: Observe context and relationships across a broad range of factors, consensus building, and general knowledge of a wide range of topics so they are able to recognize patterns of how things may relate or interact.

- Personal Connection: Will appreciate the opportunity to work across multiple topics, teams, or projects. Communicate best with them by sharing the big picture goals first as they will naturally see how the "puzzle pieces" fit together.

- Activate and Direct Potential: Most powerful coordinators, connectors, and strategic planners. May also be natural team leaders as they will be able to relate across many tasks and topics with ease. Will burn out if placed in an environment requiring in-depth, ongoing specialization on one subject for too long.

- Common Roles: Examples of those scoring higher on the complexity scale would be general physicians, operations management, project managers, general teachers, marketing leaders, and strategists.

Specialists

- Peak Contributions: In-depth knowledge of chosen subjects, subject matter expertise, risk identification, and mitigation.

- Personal Connection: Acknowledge their specialized knowledge as a valuable contribution and avoid giving ambiguous direction on topics outside of their specialty.

- Activate and Direct Potential: Will not feel comfortable being rushed to make a decision without feeling fully knowledgeable about it first. Will appreciate the opportunity to focus on topics of interest or specialty. Burnout will occur if a specialist is assigned many seemingly random tasks that are out of alignment with their specialty.

- Common Roles: Examples of those scoring lower on the complexity scale would be subject matter experts and authorities, such as a heart surgeon, software specialist, professor, or rocket-fuel scientist.

The Strategic Advantages

The AEM-Cube and its insights have been successfully deployed in more than 1,200 organizations globally. Understanding and proactively using this information has several advantages for the organization. Strategically important tasks are accomplished more effectively, with greater velocity, by more engaged and fulfilled teams of executives, to the clear advantage of all stakeholders. The value depends on whether it is applied to individuals, teams, or the organization as a whole.

Individuals typically benefit through self-awareness, personal development, and career planning. It helps them:

- Discover where their natural peak performance, strengths, and talents lie to contribute most powerfully to change

- Find the most energized way to get into action effectively and connect with a team
- Map out their leadership style and related potential roles within the organization

Teams often benefit by gaining insights about their strategic variation (diversity), execution dynamics, innovation, and performance. The AEM-Cube maps out where strengths and weaknesses lie within the team and also shows whether they are sufficiently equipped to achieve the necessary result areas. Further, it contributes to improving mutual relationships. It helps teams:

- Map interpersonal dynamics to ensure sustainable growth
- Analyze gaps and risks to ensure sustainable growth
- Nurture mutual appreciation and understanding for everyone's unique contribution to team performance
- Identify the cognitive and strategic diversity of teams by pooling individual contributions

Organizations can leverage the AEM-Cube to align teams with performance goals. Moreover, the AEM-Cube links the growth potential of individuals and teams to the Growth Curve of the entire organization. This way, innovation and change processes will run more smoothly. The tool helps companies:

- Provide insight into connecting the optimal contributions of individuals to the strategic objectives
- Link the growth potential of people to the natural Growth Curve of the organization

- Help to identify potential energy across aspects like innovation, engagement, productivity, and change processes
- Identify target areas to increase the benefits of strategic variation or cognitive diversity[20]

Whether identifying the peak potential energy for an individual, team, or organization, the AEM-Cube provides a model to uncover potential opportunities, align with strategy, and identify gaps. Using the tool has proven to be an important part of an effective Human Dynamics program. It is like a triage process that leaders can use to visually understand a company's variation across the three axes (attachment, exploration, and managing complexity) and adjust as needed.

While we have found the AEM-Cube tool helpful as an objective, quantitative tool that has research and scientific rigor behind it, there are other ways to informally identify how people get into action. In ideation sessions, for example, you can write groups of words on the board that represent the extremes on the Growth Curve (e.g., *explorer* versus *stabilizer*). Then ask each participant to choose the words with which they resonate most. This provides a simple way to design short-term teams by being sure to include representation from a mix of strengths.

Chapter 9

FROM THE TOP: EXECUTIVE LEADERSHIP DYNAMICS

The problem with being an executive is finding the time to read all the books on how to be a better executive.

—BOB NELSON

"The Board has decided to bring in another CEO to stabilize the company. I'm sorry, it's time for you to step down…"

These were not the words Harvey Bishop expected or wanted to hear. No CEO does. Five years earlier, he'd been brought in to lead Trudell, a 15-year-old, small, privately held, struggling administrative services company that assisted health insurance providers. Known for his innovative style, Harvey was able to spot opportunities quickly, with seemingly little effort. He was the kind of leader who inspired those around him to take risks and explore unchartered waters. He felt he was just getting his feet wet.

At just 30 years old, Harvey was well known around town for creating an app that allowed hospital patients to see their place in line for services in the waiting room. Two years after its launch, a software company bought the app, making Harvey a wealthy man and giving him the opportunity to meet other, more experienced business owners in the region. A tall, fit man with a nervous smile, Harvey was always devouring periodicals and new books, driving the latest car, and moving quickly between his volunteer roles, work, and home life with dizzying speed. He was recruited for the CEO position by an acquaintance from the golf club who thought Harvey would be a great fit; an entrepreneur with the proven ability to create a successful startup company in the healthcare-tech sector.

During his time at the helm, Harvey breathed new life into the tired company. He was onboard just six months when he introduced a new process that revolutionized the organization's current practices. Within 18 months, the formerly stagnant call center was teeming with activity. Harvey grew revenue to 10 times what it had been and added over 500 new jobs to the local economy. In fact, they had to move into a new facility to accommodate the company's high-growth expansion needs. With each passing month, Harvey continued to see new possibilities and quickly deployed them, back-to-back, constantly reinventing.

Things were going great…until they weren't. By the end of Harvey's third year, two of the original call center executives had left. The IT director, Rob, said he couldn't keep up with all the system changes. He warned Harvey on many occasions that "doing too much, too fast, will create instability," but his concerns fell on deaf ears. Harvey rationalized Rob's departure by telling himself the IT director was resistant to change. Similarly, after leading call center operations for eight years, Karen took another job, claiming she could not provide quality training for customer service representatives as quickly as Harvey was rearranging the features offered. "She just doesn't get it," he told himself. He replaced both Rob and Karen with like-minded executives who were eager to focus on the growth initiatives upon which Harvey thrived.

Instead of continuing along the same growth trajectory Harvey and Trudell had enjoyed for the first few years, the company lost focus. System glitches proliferated and compromised relations with insurance providers, customers received inconsistent answers from different call center reps, and hospital administration service providers began to feel a sense of chaos from the constant influx of changing operating procedures. Instead of streamlining and scaling, Harvey kept coming up with the next big idea, and chasing "the shiny objects."

Harvey was like a speedboat zipping across a smooth and tranquil lake. All people who thrive at the very beginning of the Growth Curve, who live in the future, and who naturally excel at producing new ideas faster than multiplying rabbits are like this, which is great for the thrill of speeding across the lake and moving quickly between one thing to the next. Speedboats are perfect when you need "disruption," a dynamic term often used when we disrupt an industry with innovation. But innovators like Harvey can also create whiplash, literally. They're at the beginning of the Growth Curve and with every major change in direction, imagine them whipping the S-curve back and forth—and those at the TAIL of it, the stabilizers and efficiency makers, are slung the hardest, the furthest, and feel the most exhausted.

When Harvey arrived, Trudell was a settled, mature company. It was like a houseboat or a pontoon. Its veteran executive team was focused on fishing, swimming, and enjoying the peacefulness and predictability of the lake. The wake (waves) from Harvey's speedboat rattled, shook up, and displaced everyone around him. Similar dynamic effects are the energy/waves that ripple out when you drop a stone/change into a lake. The bigger the change, the higher and more disruptive the waves are.

Ultimately, **the very attributes that made Harvey such a success as CEO and that led to Trudell's incredible transformation were the exact same qualities that prevented the company from scaling, maturing, and stabilizing.** He treated the company like a startup, which was needed at first, but he never eased up on the throttle or allowed the waves to subside. He was out of tune with the dynamics of the other people on his team, and unaware that his strengths were overpowering his own good intentions. If Harvey would have understood the need for a shift in company priorities as the business entered new phases of the Growth Curve, perhaps he and the organization could have had a more successful outcome.

AAA: Assess, Align, and Adapt

At the executive level of leadership, companies need the right mix of attachment, exploration, and complexity—the three axes measured by the AEM-Cube—to mitigate risks and maximize opportunities. Growing successfully in a chaotic, unpredictable environment requires dynamically changing the team at the top to align with the current and upcoming phases of the Growth Curve. This can be accomplished through the Human Dynamics Framework of Assess, Align, and Adapt (AAA).

- Assess—Companies must be clear about their current and future life cycle stages to maximize opportunities and mitigate risks during that phase. For individuals, this means understanding how they are energized most to contribute at peak performance.

- Align—Executive leadership must align key contribution preferences with the needs of the current and upcoming growth phases.

- Adapt—Once companies know what to do, when, and why, based upon assessment and alignment, they must consistently adapt to shifting opportunities and threats.

Executive leadership is not one-size-fits-all, and no leader can actively contribute to every phase along the Growth Curve with equal potential and power.

Harvey's experience as CEO at Trudell attests to this, as does the Southwest Airlines Christmas 2022 Meltdown. The company made worldwide headlines when a major winter storm hit during the peak holiday travel weeks, stranding a record number of holiday passengers. Much of the blame for this disaster has been assigned to Southwest's network operating model, but there's much more to the story than that.

The root cause of Southwest's meltdown stems from an issue directly related to Human Dynamics and the necessity to Assess, Align, and Adapt when there is an imbalance between the executive leadership's core strengths and the company's current phase along the Growth Curve.

Gary Kelly is a highly qualified and acclaimed CEO who led the airline giant to new heights from 2004 to 2022. The company was one of Wall Street's biggest success stories when it hit peak Maturity around 2017. The following chart depicts Southwest Airlines' stock price as it fluctuated for over 40 years.

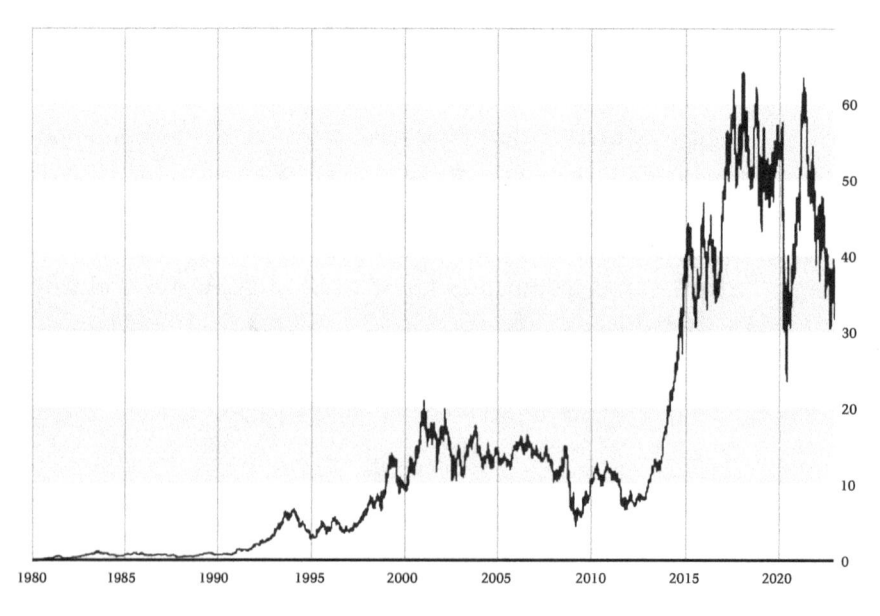

Source: Southwest Airlines

Note that Kelly's optimization preferences and specialized knowledge of financial efficiencies were ideally leveraged as the company emerged into its Growth Phase from 2011 to 2015. His strengths helped the airline scale and become more specialized in operational efficiencies, achieving rapid increase in shareholder returns during the process.

The company's net income is shown in the next chart, beginning in 2010. Notice the bell curve shape, with the Maturity Phase beginning in 2015. Think of the Maturity Phase as being a "cash cow," where current operational focus is on high efficiency and low risk: let the bovine graze in its pasture until every last blade of grass has been devoured for nourishment. Long before an organization reaches Maturity, it must be thinking strategically about its inevitable Decline.

SOUTHWEST AIRLINES CO.'S NET INCOME FY 2010 TO FY 2021
(IN MILLIONS OF U.S. DOLLARS)

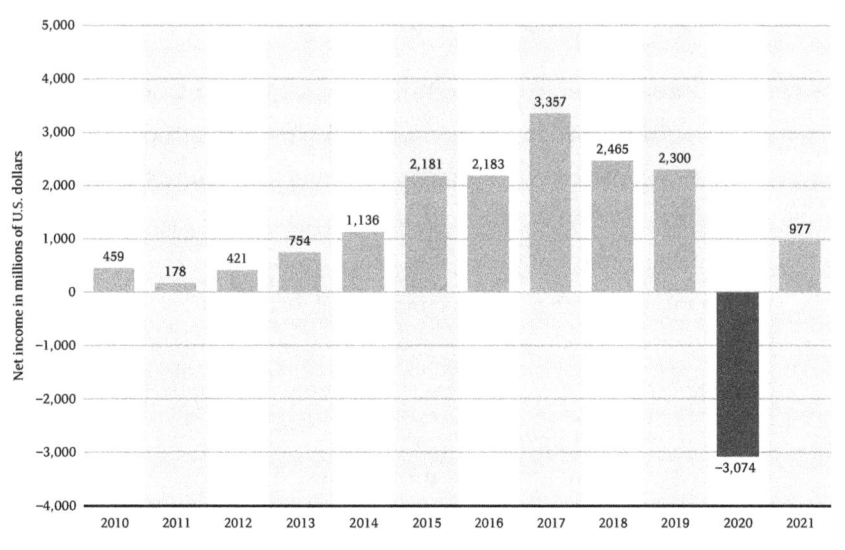

Year	Net Income
2010	459
2011	178
2012	421
2013	754
2014	1,136
2015	2,181
2016	2,183
2017	3,357
2018	2,465
2019	2,300
2020	−3,074
2021	977

Source: Southwest Airlines

Even though the Covid-19 pandemic was an anomaly that caused steep declines during 2020 for the travel industry, Southwest was unprepared to reinvent itself afterward. It continued using financial gymnastics to support stock prices, rather than assessing its situation and aligning its Rebirth strategy. When Southwest brought in a new CEO in 2022, their choice did not provide the company with enough variation from its current leadership focus.

The AAA Framework as related to Southwest, in a nutshell:

- Assess—Southwest was well within the Maturity Phase with signs of entering Decline, so executive leadership's strategy should have focused on finding an executive who was a natural change agent or innovator to help the airline identify, then leap onto, the next S-curve.

- Align—Southwest did not shift executive leadership to align with its current and upcoming phases of the Growth Curve until it was too late. Ideally, they would have changed CEOs around 2017, at peak Maturity.

- Adapt—When Southwest did finally bring in a new CEO in 2022, Bob Jordan, they chose a 34-year Southwest veteran whose contributing strengths were nearly identical to those of Gary Kelly: optimization, efficiency, and risk reduction. Jordan began his career as a programmer, and then served in Finance, Planning, Procurement, Fuel, Facilities, and Technology. While having a wider range of complexity across a number of specializations, Jordan was still deeply entrenched in the current leadership paradigms. Instead of igniting a new spark, he accelerated the airline's eminent Decline.

Southwest missed the opportunity to leverage organizational leadership dynamics to prepare for and adapt to Rebirth, as illustrated in the figure that follows.

S-CURVES IN ACTION

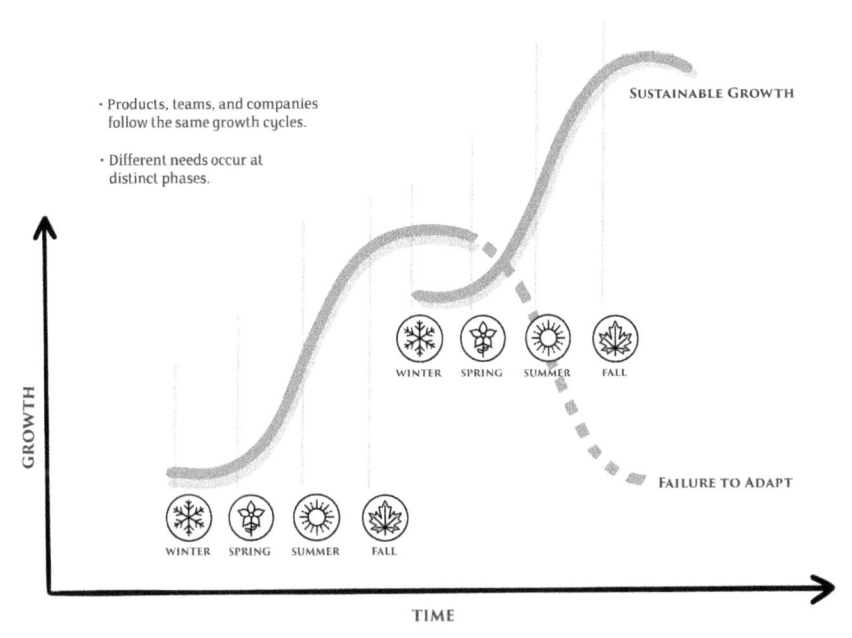

- Products, teams, and companies follow the same growth cycles.
- Different needs occur at distinct phases.

SUSTAINABLE GROWTH

WINTER SPRING SUMMER FALL

FAILURE TO ADAPT

WINTER SPRING SUMMER FALL

GROWTH

TIME

Source: Developed from Human Insight chart.

Without Rebirth, the same business practices that made scalability and specialization thrive during the Growth Phase ultimately cannibalized the organization during Decline.

Missing the Leap: An Executive Alignment Analysis

Prior to recent missed opportunities, Southwest's organizational leadership dynamics have historically been well aligned with its position on the Growth Curve. Herb Kelleher, Southwest's co-founder, led the airline until 2001 and contributed powerfully during the Birth and early Growth Phases. His legacy was built upon exploration of new business models, a relentless focus on customer and employee satisfaction, and a high degree of complexity.

Kelleher was energized by working with people. He viewed

shareholder value as the output rather than the primary goal, and he demonstrated this passion by embracing customer and employee feedback and improvements on a continuous basis. Kelleher said, "*You put your employees first. If you truly treat your employees that way, they will treat your customers well, your customers will come back, and that's what makes your shareholders happy.*"

Beyond exploration and attachment, Kelleher was a generalist who was comfortable managing complex decision-making in a decentralized way, putting the power in the hands of frontline employees and specialists. He kept a constant "ear to the ground" for ways to reduce friction in processes and to establish, then build a customer-centric brand. Leaders who are natural explorers take risks, encourage innovation, and enjoy change.

In many ways, Gary Kelly's contribution strengths and peak potential were complementary to Kelleher's, which made him an ideal leader to transition from the Growth to the Maturity Phase of the Growth Curve. Unlike Kelleher, Kelly was more of an optimizer—energized by finding and reducing risks. It's no surprise he began his career as a CPA who was an audit manager for Arthur Young & Co. and controller of Sterling Software. At Southwest, he served in various financial roles, including CFO, before landing the chief executive role.

Based on the priorities Kelly implemented, his attachment preferences appear to be on the side of content, meaning he was most energized by working with processes, financials, and systems instead of drawing peak energy from collaborating with other people and customers. Kelly was also more specialist than Kelleher, focused deeply on stock price and shareholder value versus internal employees. This shift was evidenced by his decision to buy back Southwest's own shares, using $8.5 billion of excess cash. However, the big picture, such as the need to reinvest in new markets or modernize technologies, was missed.

A high-level analysis of Kelly's tenure, and specifically his key contributions compared to Southwest's Growth Curve, can be glimpsed here:

- Maturity Phase

    - Operations (mostly aligned): Refine business model, improve operational efficiencies, establish streamlined processes, maximize profitability, protect brand, identify and guard against risks. Actively seek innovation (this aspect lagged).

    - Strategy (not aligned): Refine rebirth or exit strategies and seek leaders that align with reinvention

    - Kelly's leadership continued "cash cow" thinking, with a focus on optimization, content, and specialization instead of exploration, human design, and generalization.

- Decline Phase

    - Operations (not aligned): Determine feasibility for continuity, minimize risks, leaders who can invigorate (innovation, sales, marketing, etc.) or divest (mergers, acquisitions, etc.). Avoid risks on existing business and begin taking risks for rejuvenation.

    - While the risk profile aligned under Kelly and Jordan's leadership, a significant emphasis on innovation and Rebirth was missed. Of note, Southwest's last acquisition was AirTran Airways in 2010.

    - Strategy (not aligned): Activate an exit strategy that will restart the Growth Curve (to continue) or that will divest (eliminate the organization). Take risks for next venture.

This misalignment highlights the reason for Southwest's undoing during December 2022. **Instead of welcoming a new strategy that would invigorate the airline afresh, Kelly and the board members passed the baton to someone with similar strengths, lacking the variation that aligned to the company's current growth context.**

Critically, Southwest was no longer connected to its employees, which happened during the switch from Kelleher to Kelly in the early 2000s. Like any network in nature, without connection, energy cannot flow productively. The culture that sparked differentiation from the airline's competitors morphed into something completely different. In pursuit of shareholder returns, instead of a natural balance, Southwest missed the boat when navigating its core energy source, Growth Curve alignment, strategic variation, and successful networks.

According to the American Enterprise Institute, 88% of the Fortune 500 firms that existed in 1955 are gone. They either went bankrupt, merged, or still exist but have fallen from the top 500 category. Other well-known examples where executive leadership failed to adapt between the Decline and Rebirth Phases include Polaroid, Toys R Us, Borders, Compaq, and Kodak. Myriad articles cite a lack of innovation as the catalyst that caused these companies to fail. While improving innovation addresses the "how" to support continual growth, it falls short of explaining the "when," "what," and "why." These organizations all failed because they did not properly Assess, Align, and Adapt based upon the current context of their Growth Curves, missing critical signals like Southwest did in this case study to transition successfully into the next phase of sustainable growth.

As a frequent flier and fan of Southwest, I hope the company uses the holiday fiasco as a catalyst for rapid renewal through Rebirth. There is still time to avoid the demise experienced by other mega-companies as they, too, missed the opportunity to plan for Rebirth at the appropriate time.

Making the Leap: Strategic Growth Curve Navigation

Successful companies are adept at aligning their CEO's strengths and strategic actions with their current and upcoming phases of growth. We already discussed Amazon's success in this area in Chapter 5, and Apple is another strong example, primarily because it is consistently energized

FROM THE TOP: EXECUTIVE LEADERSHIP DYNAMICS

toward Rebirth and its executive leadership dynamics reflect as much.

Apple was originally born in 1977 as a computer company. When the first version (Apple I) reached Maturity, the Apple founders and innovators continued to roll out newer and faster versions of desktop computers. From there, Apple successfully leapt onto multiple new S-curves with a variety of novel and proprietary digital devices. Those individual products formed small S-curves that kept Apple in a continual state of Rebirth with the iPod and iTunes in 2001, the iPhone in 2007, and the iPad in 2010. The smaller life cycles combined allowed Apple to stretch out its Growth Phase for the next four decades, as shown in the following figure.

SUCCESSFUL ADAPTATION

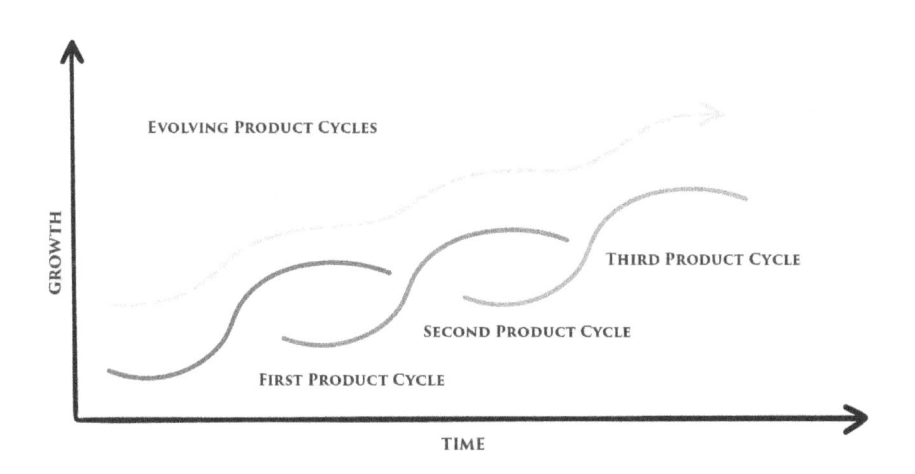

Source: *The Physics of Life*, A. Bejan*
*Note: I varied this diagram slightly to show successive curves beginning between the growth and maturity phases instead of at the beginning of the decline phase to leverage natural momentum and reduce exertion needed for each subsequent startup.

Beyond consistent and groundbreaking product innovation, Apple's leadership dynamics fluctuated at several important junctures in the company history. Contrary to popular belief, Steve Jobs did not become the CEO of Apple until 1997. He was a co-founder with

Steve Wozniak, but there were four other men who held the CEO title ahead of him, during the company's critical Growth trajectory in the '80s and early '90s: Mike Markulla (1981-83), John Sculley (1983-93), Michael Spindler (1993-96), and Gil Amelio (1996-97). Each of these men had their own effects on the company, some highly positive (Markulla, attributed with doing more for the company than anyone else, except for Steve Wozniak himself, who invented the first two computers), others less so (Sculley was ranked by Conde Nast as the 14th worst CEO of all time).

After Steve Jobs was famously fired in 1985 from the company he helped to create, the Board brought him back (at his own suggestion) as "interim CEO" in 1997. The company was in active Decline, with stock prices at a 12-year low, and his sole purpose was to make the company profitable again. Jobs famously had little regard for people and the politics associated with managing them. Employees feared his "my way or the highway" leadership style and for good reason. He cut lagging products and the teams behind them with precision. Using his style of targeted innovation, Jobs leapfrogged onto the next the S-curve, and transformed the music industry with the iPod rollout. And what did he do next? He cannibalized iPod sales by creating the iPhone. "If we don't cannibalize ourselves, someone else will."

In 2011, when Jobs stepped down due to illness, he had taken the company from one phase of Maturity into an entirely new Growth Curve and back up to Maturity again. When Tim Cook took over in Jobs' wake, the executive leadership dynamics at Apple shifted dramatically once again. Faced with an almost identical business challenge as Jobs had faced when he came on board, and rather large shoes to fill, Cook's energetic focus and corporate competencies could not be more different than his predecessor's.

Prior to the CEO role, Cook served as COO with known focus on driving profitability and crafting complex global supply chains. Compared to Jobs, he is "not a product person," but he is outspoken about the responsibilities that accompany ownership of the products his company creates. He believes in collaboration, transparency, diversity (cognitive in particular), and trust. His modern-day leadership attributes have provided the company with the stability associated with Maturity, and he has grown Apple into a multitrillion-dollar company.

No doubt, Apple and its executive leadership team are already considering what's next. New iterations of the iPhone can only sustain the company's position for so long before it will need to reinvent itself. However, with reliable confidence, the company has demonstrated a decades-long practice of leaping onto the next Growth Curve through constant innovation and adaptation in both its products and its C-Suite choices.

Apple has exemplified that growing successfully in a chaotic, unpredictable environment requires Assessment, Alignment, and Adaptation to ensure the right mix of organizational leadership dynamics for the current and upcoming growth phases. Those executive-level dynamics set the tone for the whole rest of the operation. **How well the C-Suite is aligned to the company's precise needs in the moment is a dynamic that is felt by everyone else around them.**

Like a towering redwood tree in the heart of a dense forest, CEOs sway in response to the changing breezes. These executive changes and ongoing dynamics at the highest levels create a resonance that reverberates through the entire organizational ecosystem. The redwood's canopy, akin to the C-Suite, captures the value from exposure and direction at the top, filtering down sunlight and nutrients to sustain the entire tree.

This alignment, or lack thereof, is mirrored by the forest floor. From the scurrying creatures to the intricate network of fungi, the forest's inhabitants react to the redwood's presence, adapting their behaviors and interactions. Just as the redwood's influence extends to the farthest corners of the forest, the resonance of an executive's energy ripples through an organization, reaching stakeholders, partners, and even the broader business landscape, powerfully shaping the culture and success of the entity.

The energy at the top is contagious, and it ricochets through the organizational culture like an invisible current that is not often seen, but always felt.

Chapter 10

TEAM LEADERSHIP DYNAMICS

*I am not the smartest fellow in the world,
but I can sure pick smart colleagues.
Or, as I like to call them, my human crutches.*

—FRANKLIN D. ROOSEVELT

Like Ted Lasso after winning the World Cup, Alfred was embraced and celebrated by his team. Tears of compassion and appreciation flowed toward the guy who was widely considered the curmudgeon of the group. Resentment dripped away as the team was finally given insight into why Alfred was such a pain in the neck. The emotional scene was not what you would expect at a Fortune 200 company in the heart of Boston; it was more like the end of a Disney movie.

Just a few hours earlier, most of the Northeast Sales Team sat together, chatting about their families, dinner the evening before, and current events at work. Alfred attempted to nod along but he looked lonely and visibly uncomfortable among the others. He was clearly an outlier in their midst and struggled to partake in chatty banter, never looking anyone directly in the eye.

Alfred: The Pain in the Neck/Pack Protector

My colleague Trent Strobel and I were in Boston to present a workshop for the sales leadership teams of a global consumer products manufacturer. Before arriving, we assessed every middle leader in the region, from managers up to senior directors, using our Assess, Align, and Adapt Framework in conjunction with the AEM Cube. We then graphed the needs and goals of each functional team to see how its members aligned, identifying gaps and risks along the way. Speaking to an audience of around 50 leaders containing six teams, Trent and I unveiled the quantitative and qualitative reasons why and how Human Dynamics (diverse collaboration, equality, and inclusion based on how each person is naturally energized) helps teams to improve. We decided to use the Northeast Sales Team as our primary example.

Everyone on the 10-person team was highly complex (energized by the big picture), had a strong preference for interaction with people instead of content, and was excited by change. Everyone, that is, except Alfred. He was as opposite as you could get on all three AEM Cube measurements: exploration, attachment, and complexity.

When the team's leadership matrix graph lit up the screen, nine tightly clustered dots stood in strong contrast to the one lone dot at the other end. The pattern on the screen echoed the emotional context we witnessed throughout the workshop. Trent and I explained, in academic terms, how functional teams must understand their unique mix of potential energy to identify risks and opportunities for optimal performance, but it was the dog-walker story that forever changed the Northeast Team's dynamics.

I explained that the nine team members who were energized in very similar ways by exploration were like dogs from the same doggy daycare. They were excited to see each other every day and they were always interested in sniffing out something new together. Sometimes it involved chasing squirrels or making a new friend in the park. Other times, it was being curious about the colorful coy fish in the pond or frolicking in the water fountain. As great "generalists," these nine doggies missed nothing, seemingly seeing everything simultaneously. They were limited only by the time available to sniff out new possibilities, and the leashes that tied them to their walker. Great inventors, like Thomas Edison, share these same characteristics. They churned out new ideas faster than a long-tailed cat in a room full of rocking chairs. The team laughed, imagining themselves as a pack of playful pups, and their pride at being compared to one of the greatest innovators of all time was palpable.

As the story continued, however, the giggles faded. "Completely unaware of the risks, one of the dogs broke free of her leash and ran at full speed to chase a bright red ball into the street. She was immediately stuck and killed by an oncoming car. One by one, the other dogs followed their friend, focused on the bright red ball bouncing between cars in the busy byway like a pinball. The dogs were so engaged in the game, the wailing warnings of their walker were completely ignored." The room became uncomfortably quiet as they absorbed the story's turn.

Then, I pointed to the single dot on the graph. That one solitary dot was Alfred, floating at the opposing end of the Exploration Axis. Unlike the rest of the pack of Explorers who were also Generalists and People-focused, he was an Optimizer, a deep Specialist, and preferred Content instead of people. He was different in every way than his team as shown in the team graph that follows. Unlike the others, he represented the dog walker.

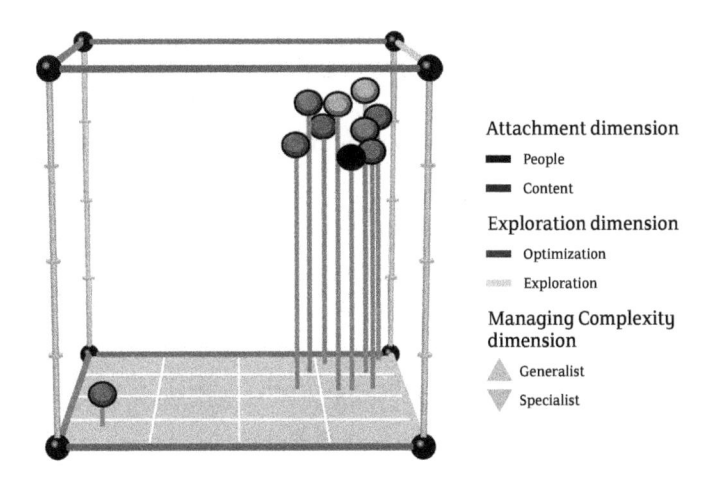

Attachment dimension
- People
- Content

Exploration dimension
- Optimization
- Exploration

Managing Complexity dimension
- Generalist
- Specialist

Alfred was constantly aware of the risks; he knew the strength of the leash held firmly in his grasp and how important it was to keep it well in hand to protect his pack. He kept away from the area of the park that ran adjacent to the busy street, always double-checked constraints to prevent an escape, and hoped that his own sense of security would help keep the pups safe. More like Henry Ford, who was annoyed by the barrage of seemingly pointless new ideas and risky changes from his top salesmen, Alfred preferred to trust what he knew to be true. The dogs, in contrast, were eager Explorers like Thomas Edison, so this type of stabilization felt like being held back from progress—or a royal pain in the neck.

For the first time all day, Alfred cautiously made eye contact with me. I smiled at him as he stood up to face his team, mustering all the energy he could while a tear slipped down his cheek. Fists clenched in resolve, and in a firm but quivering voice, he said, "I never wanted to hold you all back. I just didn't want you to make a mistake that could hurt us all."

I knew how much courage it took for Alfred to say these words to his team. Prior to the workshop, I had contacted Alfred individually. When I saw his and his teammates' AEM Cube results, I wanted to get his permission to use his profile as an example for the whole group. I also wanted to get a better understanding of what the dynamics were like from his unique perspective, as the only Specialist on his team.

During our call, he shared that, as a kid, he'd been bounced between extreme learning environments; from special ed to gifted classes. He never made sense of his skills and challenges until his own son was diagnosed with autism. Upon learning more about the developmental disorder, he recognized many of the attributes in himself. In fact, he was hired to be on the sales team because he knew every minute detail about each of the consumer products the company manufactured. They weren't just a job to him; they were part of his everyday life, and he took his position very seriously.

When Charlotte, the team lead, heard Alfred's words, she leapt up out of her seat and said, "Is it okay if I give you a hug?" He nodded shyly in affirmation and seemed to collapse a little with relief as, one by one, the whole team stood to thank him for looking out for them.

Trent and I didn't expect such a profound emotional awakening. Theirs was an interaction of equity, respect, and inclusion that demonstrated the change each of us can be in the world. Alfred had finally been seen and valued for his contribution to the team.

Intra-Team Dynamics: Synergistic Power

Intra-team balance is derived from the right mix of variation and inclusion to maximize the potential for power, speed, and agility within each phase of a project. Getting it right creates synergy. In physics, as in business, synergy is a scenario in which greater value is achieved by group dynamics versus individual efforts, or, as Aristotle said, "The whole is greater than the sum of its parts." **Continuous variation and adaptability to changing pressures, needs, and Growth Curves will always outperform traditional long-term, fixed goal-setting processes in non-linear business.** Gone are the days of forming static, inflexible teams in a one-size-fits-all approach.

Optimizing intra-team dynamics and composition depends upon the specific context at hand, which includes the project's propensity for change, complexity, and orientation to people or content. Team leaders must consider key attributes each project requires before determining who should be on the team.

> Q: Will it require the generation of new ideas or refinement of an existing process?
> Q: Is it customer facing, behind-the-scenes, or a mix of both?
> Q: Will it have an impact on other teams, processes, or stakeholders; or is it an isolated event that needs deep expertise?

The process of identifying a project's key attributes informs the team's needs and allows for the consideration of a custom mix of exploration, attachment, and complexity contributors.

The more specialized a team's function, the more skewed its groupings should be to maximize direct power output and minimize line loss. The marketing, sales, and innovation teams, for example, maximize success when leader contributions are skewed toward the beginning phases of the Growth Curve and more generalist, whereas engineering, accounting, and compliance perform best when contributions are skewed toward the later phases.

Team leaders must also select team members that align with anticipated needs for the project. Every person is energized in different ways and is most connected, engaged, and productive when their role aligns with what naturally stimulates them into action. In more traditional companies, this process often works backward—team members are fixed and the project changes. In Human Dynamics, however, the project needs are assessed first, then the people vary to align and adapt to the current project.

In an environment of accelerating change, past projects are not necessarily indicators of future success; only the current is relevant.

While people will likely need to serve in ways that are outside their natural contribution preferences on occasion, team leaders must recognize that the more extreme an employee's preferences are, the faster and higher their risk of burnout. In addition, the less aligned team members are to their duties, the lower their engagement and more likely they will experience feelings of isolation and loneliness. EY's survey of more than 5,000 workers in Brazil, China, Germany, the U.K., and the U.S. found that *82% of respondents indicated they have felt lonely* at work, so it is important teams understand and acknowledge individual contributions.

Team Dynamics in Action: From Flop to Fortune

Just before the onset of the Covid-19 pandemic, a Fortune 500 retailer used the AEM-Cube to assess its IT leadership teams. We mapped existing teams, identifying risks and opportunities within each, and held workshops to inform 97 leaders about the importance of Human Dynamics. Little did we anticipate what was coming—the pandemic would accelerate the retailer's demand for integrated technologies by three years, which created a mountain of immediate needs.

The retailer was faced with the challenge of delivering the same quality of service to their customers online as they had in real life. Curbside delivery and customer interaction, which normally occurred in-store, was integrated into a new, custom-designed mobile application. The Applications Team rushed into action to meet the new demands. Due to necessity, they created a highly functioning app and rolled it out to customers everywhere after only abbreviated testing. Nobody stopped to Assess project needs, Align team members, or Adapt in a way that would optimize individual contributions.

A few months after the new app was released, users made their voices heard by awarding it a one-star rating. The negative comments were mostly related to poor design and integration, like it was not intuitive or simple to get through the checkout process. Also, customers could order products from the app, but once they arrived to pick them up, there was no way to notify the store without going inside.

These results were unexpected considering the expertise and capabilities of their Applications Team. Known as a very customer-forward company, the retailer was determined to make it better. They started with the question: How could this have happened? Even though capabilities and expertise were up to par within the IT team, it consisted of many like-minded individuals who shared the same profiles, backgrounds, and thinking patterns.

As indicated in the Applications Team graphs shown here, this team contained members who were mostly energized by finding risks and working with content. In addition, they engaged best by performing highly specialized roles.

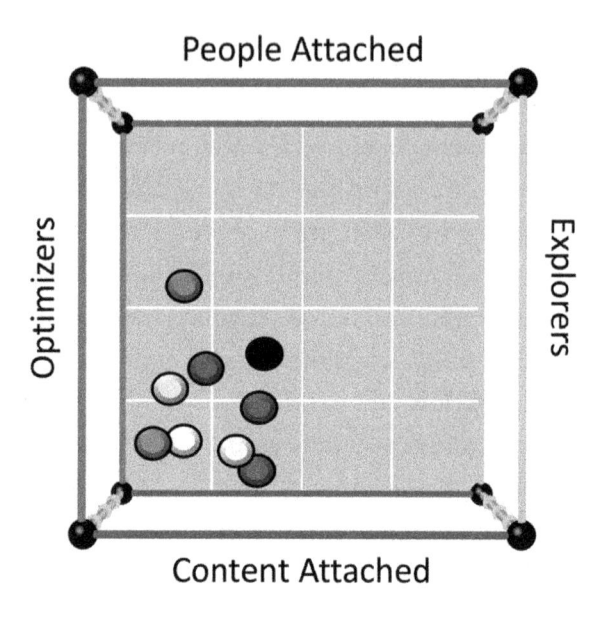

People Attached / Optimizers / Explorers / Content Attached

When asked to find a way to quickly improve the new customer app's user ratings, we started by using the first tenant of the AAA Framework to take a step back and Assess the project's needs. We asked the following three questions:

> Q: Will the project require generation of new ideas or refinement of an existing process?
> A: The customer app required new ideas since it was to be custom-built to meet new needs caused by the pandemic, so there was no frame of reference. This team needed Explorers who naturally brought innovative solutions that had not been tried before.

Q: Is the project customer facing, behind-the-scenes, or a mix of both?
A: The app required customer-facing expertise, or those who naturally empathize with people, to design an intuitive customer journey. It also required content experts who understood the technology necessary to perform in a robust way behind the scenes.

Q: Will it have an impact on other teams, processes, or stakeholders; or is it an isolated event that needs deep focus?
A: The new app needed team members with a high degree of complexity, meaning they could spot how different processes and systems would interplay most effectively.

After Assessing the needs, we all agreed the like-type thinking of the Application Team seemed to have contributed to blind spots about customer experience and designing new ways of interacting that had not been done before. To Align this project team to the task at hand, the IT Department Adapted by pulling in tech architects from the enterprise team as well as members of the marketing team and others, like store managers, who were energized by working with customers.

These changes yielded rapid results. Insights were gleaned from users, new journey maps were created, and design thinking was used to ideate, prototype, and rapidly test new versions of the app. Within two months, user ratings went from one star to 4.6 stars and app-related revenue increased by 180%. Providing new updates to the technology helped the retailer's customers order from anywhere and pick up their needs safely from their vehicles, which was critical during the peak of pandemic uncertainty.

The challenge the Application Team originally faced was that they did not have the benefit of cognitive diversity. Breathing new life into any project is not a mystery; it's a matter of finding and deploying the right mix of energy. **Once a leader understands what energizes their team members most, they can tap into the under-used potential to create or regain success.**

The AAA Framework as related to this Fortune 500 retailer, in a nutshell:

- Assess—The Apps Team members were overwhelmingly risk averse, yet they were tasked with inventing something that was custom and new. They were also content focused when the app needed design inspiration and empathy from customers.

- Align—The team did not shift its members to align with project needs until after the initial product failed. Ideally, they would have initiated product design and development by including a more diverse team at the onset.

- Adapt—Once the Apps Team brought the optimal mix to the table based on the context at hand, it quickly improved customer ratings for the new Customer App, helping support the company's highest growth rate (over 30%) in its over 60-year history.

Inter-Team Dynamics: Flowing Synergy vs. Isolated Silos

In the pre-digital age, businesses created separate work areas to streamline operations and master specific skills. Lines of business were like building strongholds of expertise. But here's the twist—those very strongholds that were meant to safeguard knowledge and efficiency can sometimes start to feel like they're holding us back. Instead of guiderails, too much specialization can limit our moves, stifle our creative spark, and keep us from benefiting from fresh ideas spread across different areas.

Think about it like this: when teams operate in isolated bubbles, it's like they're in their own worlds. This separation can be a roadblock to quick changes and adaptations. Traditional structural setup is like dams built by clever beavers to block the natural flow of things. While it better serves the need of the beaver's family, it impedes nourishment from reaching others downstream.

What does that mean in corporate speak? It means less productivity, fewer game-changing ideas, and a dip in the factors that keep the company moving forward—output. So, as you navigate the ever-changing waters of business, remember this: embracing a more fluid, interconnected approach can set the stage for a power surge. It's like opening up the floodgates, letting energy flow freely between teams. That means higher productivity, more innovation, and a stronger output. It's like turning those rigid structures into free-flowing rivers where ideas and creativity move freely, carrying your business toward success.

Silos also occur during the latter part of the Growth Phase when organizations typically create specialized functions to support scaling up across geographies and markets. While startup leadership teams often wear many hats, the further an organization climbs up the Growth Curve, the more siloed it usually becomes. Investopedia defines company "silos" as a metaphor for separate entities that stockpile information and effectively seal it in, like our proverbial beaver trapping all the good stuff from the stream for his own family. Further, organizational silos are business divisions that operate independently and often avoid sharing information. These **inflexible structures may feel more like straightjackets as they prevent ideas and data from moving around freely.**

According to Forrester, *some 72% of firms* say managing data silos across multiple systems, technologies, and regions is moderately to extremely challenging. Although specialization may be necessary to achieve efficiencies, to leverage the power sourced by other teams, specialized teams must find a way to allow a fluid flow of ideas and data across their web of connectedness. And, in an environment of accelerating change—where technological advances or unexpected disruptions on social media can activate chaos in mere minutes—the restricted, isolated thinking within specialized teams wastes precious time and prevents synergy. The energy dynamics of operating in walled-off silos means circuitry can't be directly connected, which prevents exchange between teams.

In systems theory, the study of cohesive groups of interrelated and interdependent parts, every system has some type of boundaries, is influenced by its context, defined by its structure, function and role, and expressed through its relations with other systems. Teams can plug into or provide power, based on the context at hand, to boost both intra- and inter-team connectivity. This allows *flowing with* change instead of reacting to it.

A captivating illustration from the natural world is the remarkable longevity of Giant Sequoias, which thrive for thousands of years despite their seemingly small, shallow roots—extending merely 6 to 20 feet deep. The secret lies in the Sequoias' interconnectedness and mutual support. Growing in close proximity to each other, their roots become intricately intertwined, forming a web of support that provides unparalleled strength and resilience. This serves as a poignant analogy for teams working cohesively, drawing upon each other's knowledge and support to build a stronger, adaptable, and more enduring system for sustained growth. Just as these majestic trees stand tall and thrive through collective strength, so too does an empowered and interconnected team flourish, emphasizing that the foundation of success lies in fostering a community of mutual support and collaboration, underscoring that no individual can truly thrive without the synergy of others.

In the past, companies used closed systems to support linear business needs through direct cause-and-effect relationships between the initial condition and the final state of the system. (For example, when a laptop's "on" switch is pressed, the system powers up.) Everything progressed neatly through predetermined processes, so management's role was to identify exceptions and correct them. This model is still effective for linear types of processes such as conducting audits, mechanics, or use of automation and bots.

The modern business environment today, however, also contains many open systems, meaning an end state can be achieved by using many potential means. Open systems, such as social media, operate quite differently than closed systems. The term "equifinality" describes this principle of having multiple ways of achieving the same outcome with its roots in developmental biology and systems theory (e.g., there may be many equal efforts to achieve the same final outcome).

In an open system environment, the most flexible and adaptable business model is both highly decentralized and diverse because there are more inputs and increased ways to interpret them. This allows companies to ride along with changes instead of just chasing them.

Human Dynamics has been designed with equifinality in mind to better leverage the unique contributions and variations needed from teams and individuals. Across business units and departments, dozens or hundreds of these teams will be deployed simultaneously. Connected circuits among teams amplifies speed, agility, and clarity. Disconnection leads to silos and untapped energy. Teams who understand their relative resources to other teams within the organization have clearly defined capacity and therefore increased potential for meaningful collaboration.

Teams are living things, powered by human energy in both closed and open system environments. Like the mycelium network in nature that instantaneously connects different trees, plants, and seedlings so that data and value flow seamlessly back and forth in the most efficient way possible, teams are most efficiently connected in corporate cultures that encourage web-like communications (connected in all directions instead of linear).

For business teams, the workplace culture serves as the network connecting teams. Depending on the culture's strength, reliability, and flexibility, team output can be amplified or insulated from success. Constant variation and adaptability to changing pressures, needs, functions, and growth will outperform the traditional long-term, fixed goal-setting processes. Learning how to manage your team's energy will enhance its potential for success and synergy.

Chapter 11
WORKPLACE CULTURE DYNAMICS

*I always arrive late at the office,
but I make up for it by leaving early.*

—CHARLES LAMB

Dana passed her 12th interview with flying colors and accepted the VP of IT Operations position with the $13 billion cosmetics retailer. The 382-person tech department she joined was responsible for supporting over 2,000 retail stores and an emerging e-commerce business worldwide. Unlike her new peers, Dana worked in a cosmetics store for years and has always loved keeping up with the latest makeup trends. She regularly shops online for the various products she's read or heard about and has a strong appreciation for what customers want. Her personal passion for cosmetics is part of why she chose to endure 12 interviews to land this coveted position.

The company was growing faster than its technical infrastructure could manage. On occasions becoming more frequent, the sprawling legacy system would crash unexpectedly. It had been hastily patched together with modern APIs and solutions like band-aids on a new case of chickenpox. By the time one problem was covered, another one popped up.

Sparking Engagement (In Under Five Minutes)

Early into her tenure, Dana became aware that she was one of the only people in her department who actually shopped from their cosmetics company. She was an actual customer, and she'd experienced the system's problems as a customer. From engineers, to data managers, to enterprise architects, very few IT members had walked in the customer's shoes either online or inside a retail store. Instead, they were all testing for functionality and usability based on the specs sent over from marketing or the work order requests from business operations. They never had a direct interface with the products being sold.

A decade ago, when mid-managers and officers were onboarded, they were required to visit a store for orientation, but amidst rapid growth, that process ceased to exist. Over time, this disconnection created a massive problem because people were working on problems they had never experienced. It was like having a team of therapists who provided counseling services for patients they had never met.

Dana wanted everyone on the IT team to have a "customer experience" by conducting an online or in-store transaction, like the "old" days. Implementing this initiative was one of her first actions at the cosmetics company because it had the potential to personally impact hundreds of professionals. Knowing it could take as little as five minutes, everyone had the power to choose how, where, and when they wanted to participate within the year. They could stop at the point of payment or walk all the way through the returns process.

A cross-functional team designed a five-question survey for everyone to take after their customer experience. One of the questions asked employees how their daily role specifically impacted the transaction. Another asked for specific improvement opportunities. Each question was designed to get to the core of IT's true culture and customer needs. Dana was clear about the goals of the initiative and encouraged everyone to complete the transaction in a way that was meaningful to them.

One employee tried to purchase vanilla lotion but was allergic to certain additives. He discovered the website didn't list the lotion's ingredients. This insight helped uncover 27 other products that were also missing ingredients. Another employee ran into a shopping cart error when she tried to buy eyeliner. The cart kept pre-filling her billing information with an address from five years earlier when she became a rewards member. The system wouldn't let her update it to her current address. Then, because the address didn't match her credit card, the transaction failed. Upon investigating this insight further, the company found thousands of truncated transactions that stopped at a similar customer journey point.

The list of insights from the customer experience experiment exploded with over 3,800 data points and 154 system improvements during the first six months! Such a simple activity got everyone talking, comparing notes, and moving in the same direction without the need for additional budget dollars. Yet, it helped result in double the revenue from online sales the following year.

By putting themselves into the customer's shoes, the entire IT team was able to directly associate their role to the needs of the business and the customers it served. The program also ignited Human Dynamics through enhanced empathy, engagement, and connectedness throughout the company's culture.

Everyone's perspective toward their work changed when their daily actions went from "fixing a tech problem" to surfing the website, shopping the stores, and cruising the customer app. They became engaged, and it only took five minutes of their time.

Leveraging the Laws of Motion

Human Dynamics fuels more than an organization's monetary value or longevity; it sets the tone for internal culture. Ultimately, executive leadership is responsible for establishing a sustainable and adaptable environment where individuals can thrive, but it's up to everyone to contribute their unique energies in support.

Dana's IT Customer Experience story is but one small example of how using the principles of energy can get individuals into positive action and build a more engaged, productive culture from the "inside out." Her program successfully leveraged the laws of physics to connect, direct, and accelerate action, while building a more adaptable culture.

Newton's First Law of Motion states that an object at rest tends to remain at rest, or at its current speed, until acted upon by an external force. It takes effort to get people moving or to alter their existing direction and pace. It also takes a nudge to ignite engagement at any level of an organization. In Dana's case, the "external force" was the creation of her customer experience initiative, which put people into motion. Had she not required such an initiative, the team would have continued *at rest*—band-aiding the broken system together.

Once the tech team was in action, the act of participating in the customer experience initiative created momentum. People started talking about their discoveries, then began comparing notes, and then suggesting and implementing solutions. It got them engaged in the products their company made and the processes through which their customers bought them. The process that began with a lack of action and connectivity ended with positive movement and forward momentum.

Newton's Second Law of Motion states Force = Mass x Acceleration, with "Force" being the impact of an idea on a company's success. **The more impact a solution carries, the greater its resulting momentum when people get into action to support it.**

For Dana, it was paramount to reconnect the IT team with the rest of the business and with the customers it supported. Her top challenges were transforming a segmented culture into a collaborative one, then identifying and repairing the technical issues as quickly as possible. Hundreds of employees were directly involved with the cosmetic company's performance, adding to the likelihood Dana's project would get implemented. Over time, as insights continued to pour in, the number of new solutions accelerated. People *wanted* to find problems so they could be fixed. They enjoyed personally contributing to the dashboard's positive growth day after day, doing their part so their customers had a smoother experience.

As demonstrated here, companies can use these laws of motion to predict how much a particular change will propel the company forward. Consider these two factors to estimate impact:

1. "Mass" is defined as a large body of matter or a large number of people. For businesses, it relates to ideas or decisions that impact the company or its employees in a big way. Ideas with small effects will create incremental progress, but those that align with major corporate initiatives (or executive "hot buttons") will carry more weight. **The more an idea relates to business goals, the bigger its potential impact to a company's bottom line, so it adds relevance and increases action.** In Dana's case, the problem impacted both a sizable number of people and a large problem, so it was a change that had potential to yield high organizational impacts. While this may seem like common sense, many organizations still have "big idea" programs that are not tied to a specific business need, which can actually backfire when good ideas don't get implemented due to existing budget and resource limitations.

2. "Acceleration," in business terms, describes how quickly a company will adapt to change. This can be increased by positioning change in a way that motivates the type of employees at your organization.

In Dana's case, she kept the project's time and intrusiveness to a minimum, then provided visible scoreboards so everyone on the tech team could see how their contributions were creating improvements across customer experiences online and in the stores. Funneling everyone's focus also leverages the Venturi effect—getting increased energy/productivity on insights by narrowing a broad request, as Dana did, using just five survey questions.

Aligning change to sync with employees also depends on whether you're working with those who love change or who avoid risks. For example, natural innovators make up only 2.5% of the general population.[21] In high-tech corporate cultures (like an emerging tech company), the pace of change is accelerated, because there will be a larger percentage of explorers who are activated by change, as those companies will seek to hire explorers.

In low-risk cultures, however, such as the insurance, medical, or financial sector, the speed of adaptation can be slower, unless positioned properly. Those industries generally hire employees based upon their ability to identify and eliminate risks. To achieve similar results for low-risk cultures, communicate changes to team leaders and employees in a way that resonates with them. For example, instead of saying, "We have a great new opportunity to improve revenue," a more effective statement would be, "If we don't implement this, our revenue growth is at a real risk of continuing."

By putting these concepts into practice, new ideas will create the greatest potential impact if they are *aligned* with the most relevant needs and wants of the executive team.

Removing Friction from the System

Newton's Third Law of Motion states that for every action, there is an equal and opposite reaction. This is true in sports teams; it is true in one-on-one interactions, in romantic relationships, among friends, and in families. The business world is no exception. Every action causes a reaction, and those reactions can be positive, negative, or neutral. In an environment that has experienced or requires dramatic change, understanding and leveraging Newton's Third Law is paramount to the likelihood of a successful transformation, especially one that is met with resistance, or friction.

I experienced the power of Newton's Third Law and its predictability directly when I was serving as CEO of an electronic payments company. Our niche was providing military service members with automobile "allotments": a savings deduction from their monthly income was disbursed to pay their auto loans. This program helped companies that provided credit like Nissan, GE Capital, and Mitsubishi receive a near guarantee of monthly debt service for a traditionally higher-risk demographic. Military members were considered higher risk than civilian customers due to the very nature of what was required of them: sudden deployment or extended training programs with very little notice.

In the early 2000s, our Board wanted to expand payment opportunities into other areas, including housing. The Residential Communities Initiative (RCI), a type of privatized military housing, was preparing to pilot two world-class communities to breathe new life into outdated government neighborhoods. For RCI to work, the Army needed to attract private investors. Ensuring those private investors that rent payments would be made on time, regardless of training or deployments, would be a key component to the program's success.

Our company first approached the government's internal financial provider by sharing the ways our allotment company could handle the complexity of privatized rent payments more effectively than they could as the bridge between the government and private sectors. They were still using punch cards in some cases, and by the time they upgraded their systems to meet the needs of RCI and modern development companies, it would be too late. It was a highly positive solution, and one we crafted carefully after many months of research and meetings.

This approach, however, created a reaction that was every bit as big as our proposed solution but in a negative direction. Federal officials scoffed and said that giving government payroll access to any private bank would create more problems than it would solve. They said it would expose the service members to unbearable security risks. For many months, there was no progress on either side. The harder we pushed, the more they resisted.

Realizing the need for a different approach, we repositioned ourselves as an interim tech solution. They wanted to maintain control and mitigate risk. Acknowledging this dynamic, we offered to become trained in their methodologies, obtain security-level access at the same rigor as required of their employees, use their existing infrastructure as a base upon which to build, and provide flexible new systems that would give them the connection to achieve thousands of complex rent payments. The service was essentially the same, but the approach changed from pursuing a new opportunity outside the federal finance division, to identifying a way the federal provider could mitigate risks and maintain control.

Beyond the laws of motion (or in this case, resistance) we recognized that the decision-makers for the government were in the Maturity and Decline phases of the Growth Curve (more Human Dynamics at play). Their top priority was to expose risk and kill it. To acknowledge these proprieties, we adjusted our language to better suit their position along the Growth Curve. While a "new opportunity" may appeal to those in the Birth and Growth phases of the S-curve, those words evoked resistance to a threat, not resonance with our target audience. Our second approach, on the other hand, appealed to their need for stability. We intentionally used reassuring words like "support your existing structure" and "reinforce your current level of security and supervision," and it worked.

Soon after this shift, the Army issued a Request for Proposals to compete for electronic payments. Our little bank from Elizabethtown,

Kentucky secured the contract—the first ever of its kind—and we beat out major financial institutions and technology providers from New York, Boston, and DC.

> Because of the everything-ness of the energy that drives all human behavior, any energy law in nature can be applied, in the appropriate context, to boost adaptation.

Energetic Judo

There are endless applications of Human Dynamics all around us; it's the laws of nature playing out.

When my sons were young, we took martial arts together. While they lost interest in about a year, we learned a great deal about using our core and projected energy for balance, offense, and defense. The Japanese fighting technique called Judo means "the gentle way." It emphasizes winning in combat by using your opponent's weight and strength to preserve your own mental and physical energy. Judo supports the idea that mindful techniques can be more powerful than sheer force by simply going with the flow of motion.

The principles of Judo come in handy outside the ring, in personal or professional settings too. They certainly helped with government financial negotiations in the previous story. There's no way we would have "won" if we had not learned how to flow within the direction that was important to the client.

Judo principles can also be used to deflect and redirect an individual's energy, an important tool for getting to the root of problems. While your colleagues are obviously not hostile enemies, nearly every company has that person whose fears or insecurities can be interpreted as an attack. Those types of people thrive on getting others to join them in their lower vibrational (negative) state through acts of intimidation, distraction, or resistance. When attacks occur, we naturally want to yell back or defend ourselves, but that can create a "pinball" effect—where arguments go back and forth without progress. What's more, all the innocent bystanders in the ball's path will feel a negative shock wave.

The next time you encounter someone creating friction in the system, or you see it happening to someone else, do whatever it takes to *resist* the opportunity to immediately fire back. Instead, calmly acknowledge their remarks, which neutralizes the energy momentarily, then ask them for their ideas for a more positive outcome or ask them to describe what success looks like. This energetic Judo redirects their momentum away from you and back toward a different part of their brain, and you'll get one of two results:

1. A brief respite to break the tension and give you a chance to determine your next move, or
2. The person may actually identify a better way of doing things, which helps everyone.

Regardless of the outcome, you have momentarily neutralized the negativity. Since safety and security are the most important prerequisites for establishing a personal connection, keeping negativity low is of the utmost importance when building a strong, positive culture.

Clear Direction = Movement

In addition to being skilled in energetic Judo to redirect negativity, executive leadership must also set clear direction, then provide a motivator to incite movement. This will reduce resistance and increase the likelihood of positive action. We saw this both at Dana's cosmetics company with her customer experience initiative and with the government for the service member housing allotments.

Unfortunately, and unlike those two previous examples, the C-Suite often tries to get people into action with ambiguous, sweeping requests or calls to action that are so vague, they lose the power to move people. Do any of my favorites sound familiar?

- "Get involved today!"
- "We want to hear your ideas."
- "Let's do this together!"

In the 1960s, University of Toronto psychologist Gary Latham and University of Maryland psychologist Edwin Locke discovered what we now hold as truth: the establishment of a relevant and actionable clear objective is one of the easiest ways to increase performance by up to 25%.

One company that gets this concept right is the World Hunger Organization.

1. They clearly explain their purpose: this child is hungry, and he needs your help.

2. They provide relevance by showing real pictures of the children in need and providing their names.

3. They have an easy, cause-and-effect call to action: Dial this number now; pledge $2 a day, and we will feed this child.

The American Society for the Prevention of Cruelty to Animals (ASPCA) utilized a similar technique in one of their ad campaigns. After releasing a video of rescued animals with a strong and clear call to action, the organization netted $30 million in the first two years alone.

Viewers were given clear direction (call this number, send a payment) to fulfill a purpose (save these ailing puppies). Further, they knew exactly what action the organization was taking with the donations.

A relevant, actionable, and clear direction is a powerful catalyst for movement: in physics, in nature, and in the C-Suite. In business, it sets the tone for both external and internal shareholders and guides the resulting cultural momentum (or stagnation).

Observing the Organizational Heartbeat

Any schoolteacher has encountered the Observer Effect when returning to the classroom after leaving students alone for a while. Everyone pretends not to see the paper airplanes circling overhead and the giggles grind to a sudden halt while orderliness resumes. The Observer Effect is the scientific fact that observing a situation or phenomenon changes it. Observer effects are especially prominent in physics where observation and uncertainty are fundamental aspects of modern quantum mechanics. These effects are also well known in fields other than physics, such as sociology, psychology, and computer science.

Understanding the Human Dynamics of the Observer Effect can help companies identify more accurate ways to measure their organizational culture than by using self-administered surveys. In other words, it's critical that leaders understand the ways their oversight will impact the outcomes of survey data, even when hiring independent firms to administer them. While surveys are useful for showing that you're interested in worker well-being, they are not always an effective tool for measuring cultural harmony or employee satisfaction because a variety of biases stemming from the Observation Effect lead to inaccurate data.

Successful companies are flattening and becoming more decentralized to adapt to the changing expectations of individual employees. While the trend toward decentralization helps employees to feel more empowered, executives cannot make informed decisions without gathering real feedback from their leadership teams to the front lines.

However, gleaning honest intel is not always easy to do, especially given the tendency to rely on quantitative surveys that reflect the employee's natural tendencies to align their responses with their employer's (or observer's) needs and desires. According to a recent article in *Forbes*, 81% of people fake happiness at work, so it is important for executives to understand that employee surveys can have limited value. Evidence shows workers often respond how they think they *should*, not how they actually feel.

As mentioned in Chapter 3, acquiescence bias, or the tendency for workers to respond to surveys with answers they believe leaders (observers) want to hear, can skew even the best employee survey results. Based on BI Worldwide's research, less than two-thirds of employees participate in their companies' surveys and give candid answers.

What's more, according to research from People Matters, three other types of common bias are worth watching out for:

1. **Positive Negative Asymmetry (PNA) & Negative Bias**: PNA theory suggests: "People who experience negative events may be stimulated to respond more strongly than those who experience positive events" and therefore they are inclined to give disproportionate weight to the negative over the positive.

2. **Social Desirability Bias**: Under the influence of this bias, the respondent selects options or gives responses that present them in positive light in front of their bosses. Employees may deny undesirable characteristics and ascribe to socially desirable traits.

3. **Demand Characteristics**: This bias arises when people understand the purpose behind the study and alter their responses so that the study meets its desired objectives. For example, simply revealing a purpose for the survey can skew results unintentionally, such as "We are applying to be included on the Best Places to Work list this year."

There are ways to leverage Human Dynamics to hear the heartbeat of your company. While it's not possible to spend time getting to know everyone in a large organization, recognizing the limitations of traditional surveys is a step in the right direction to seeking the truth and building a foundation of trust among all individuals. Leveraging independent sources to seek the truth can be an important part of cultural transformation. So too can deploying a reputable qualitative insights and research firm with seasoned ethnographers and anthropologists who are trained to get around bias by observing both spoken and nonverbal feedback.

While quantitative data can identify the who, what, when, where, and how, only qualitative data can uncover the all-important "why."

Alignment in Action

One example of a workplace culture whose employees are vocal about getting it right is Microsoft, as the company consistently ranks among the most enviable workplace cultures in the world. According to the organization, "We will only achieve our mission if we live our culture. We start with becoming learners in all things—having a growth mindset. Then we apply that mindset to learning about our customers, being diverse and inclusive, working together as one, and—ultimately—making a difference in the world."

While no company offers a perfect culture, the employee survey process and other key practices at Microsoft align with many of the Human Dynamics, creating the environment for a positive organizational culture.

- **Being actionable and relevant:** Each day, the company conducts an opt-in survey of a random sample of 2,500 global employees on a range of topics. This gives employees the choice whether to participate or not and it focuses on current, relevant topics versus standard annual survey items.

- **Leveraging strategic variation:** Microsoft caters to a diverse range of people, and they're proactive about ensuring their employees mirror this varied customer base. This strategy enhances qualities like understanding customer needs, fostering creativity, and driving innovation.

- **Using collaborative networks:** Microsoft has been on a cloud-migration journey for years, leaning fully toward the shared collaboration and security of cloud-based tools. This reduces silos and boosts efficiency as links to shared files eliminate version control issues. They also leverage a range of apps that appeal to different needs—like Teams, Yammer, and OneDrive.

- **Providing clear direction connected to purpose:** Microsoft has implemented an Objectives & Key Results (OKRs) program where employees and teams can directly connect their personal roles to the top organizational goals. This provides clear purpose for each person as their contributions align with the company's purpose.

- **Recognizing and preventing burnout:** The company recently patented new technology that focuses on employee health and well-being. It leverages voluntary wearable devices that detect signals of elevated stress. If symptoms reach a threshold, a wellness insight is triggered. Note, it is important that employees are energized by this notion of having devices tied to their stress levels; otherwise this method could have an equal and opposite reaction if people feel they're being unduly monitored or micromanaged. The key is not a one-size-fits-all approach; it's giving them a choice, knowing your employees, and aligning to what energizes them.

In today's complex hybrid-work environment, using the laws of energy itself can help companies align with, engage, and retain high performers. A positive culture that amplifies creativity and engagement requires more than a poster on the boardroom wall or a Statement of Values on the corporate website. It is fueled by authentic, multi-directional transparency and by creating an environment that encourages employees to assess, align with, and adapt in a way that maximizes their personal and team potential. Companies that practice these Human Dynamics will shift from taking a pulse to getting the blood pumping.

Chapter 12

PURPOSE: THE FOURTH DIMENSION OF SUCCESS

The two most important days in your life are the day you are born and the day you find out why.

—MARK TWAIN

Reconnecting

"Hey, lady, are you okay? I'm Dr. George and I just wanted to check…" The sweet, older gentleman gently touched my elbow with well-intentioned concern. He clearly thought I'd gone bananas right there in the grocery store, belly-laughing with such delight that tears of joy streamed from my eyes, as I stood facing the shelves of deodorant. How could I explain this moment to him?!

I thanked him for his kindness and assured him, "I'm fine, just blown away by all the choices!" His worrisome brow furrowed further. I'm certain if we had been in the South, he would have been thinking, *Bless her heart, that woman has lost her pea-pickin' mind.*

I was sleep deprived and achy after a four-day cross-country drive tucked between the mattress topper and a giant bag of shoes. The car was chock full of whatever would fit inside for the move from Nashville to Lafayette, a little town that lay nestled in the Rocky Mountain foothills near Boulder, Colorado.

The first task after the drive was signing a year-long lease on a sight-unseen home. Luckily, it checked out, at least for the most part. The landlord informed me about the particulars of the place and showed me the three large cans in the garage for curbside pickup of trash, recycling, and compost. *Well, that's cool*, I thought, since we were accustomed to driving 20 minutes to drop off our recycling because it was a rare practice back home.

Next, I headed to secure essentials before dark at a nearby grocery store. Along the way, I passed miles of cycling and walking trails, a couple of hot yoga studios, a farm-to-table grower, and riding stables. Tucked between the New Zealand Handmade Pie shop and the Mudslinger's Pottery Studio, the little township's sign read, "*Welcome* to our Creative, Diverse, and Eclectic Town."

Upon arrival, it seemed this was a place of open spaces and open hearts where I was surrounded in every direction with things that felt energizing. I had been paddling upstream for so long to comply with the values around me, perhaps it was time to drop the oars and allow the current to do the work instead.

I stumbled into the local grocery completely unaware how frayed I was both physically and emotionally from the challenging road trip. No doubt, my exhaustion amplified the pleasant little surprises that kept popping up like Easter eggs on a Sunday morning hunt. I noticed there were no plastic bags in the store, the produce signage included the names of the farms where the items were grown, the meats were sourced locally, and when I remembered to pick up deodorant, well, it put me right over my slap-happy edge.

I'm used to finding a couple of manly bars, like Tom's cedar fragrance or maybe Schmidt's charcoal scent, in the aluminum-free deodorant section, but lo and behold! There I stood, in front of row after row of natural antiperspirants with so many brightly colored labels and beautiful scents from which to choose. I felt like I was peering at the pot of gold at the end of the stinky sweat-avoidance rainbow. My body began to relax into the certainty that this just might be exactly where I belonged.

Earlier that morning, driving across Kansas, I whizzed through the rain on the wide-open, windy plains. Struggling to keep the car on the road in the whipping winds, I suddenly heard a loud "Whack!" on the back of my vehicle. In the rearview mirror, I saw my two bikes clinging for dear life to what used to be a sturdy bike rack. The 65 mph side drafts had destroyed most of it. Without the luxury of an exit nearby, I turned on the hazard lights and pulled to the side of the road to assess the situation. Luckily, both bikes were still intact, but they wouldn't survive the rest of the journey without reinforcement.

Dad always made my sister and me fix our own bicycles, skateboards, motorcycles. If we used it and it moved, it was our responsibility. When we were old enough to own a car, one of the things he insisted on was that we always keep duct tape and bungee cords in the trunk, "just in case." *Thanks, Dad!* But the trunk was precision packed for the move. I had to slowly unpack its contents to access my survival treasures buried all the way under the floor mat. My resolve only deepened as 18-wheeler tractor trailer trucks barreled past, throwing wet grit from the roadway all over me and my things. The winds were so strong, I had to barricade and anchor my stuff underneath the vehicle to keep them from blowing away while I rummaged around.

I repacked all those soggy sacks back into the trunk and felt like MacGyver crafting together a makeshift bike rack. Using a whole role of Gorilla tape and two purple bungees, the contraption earned a few amused honks from passing truckers, which I chose to accept as affirmation of my handiwork. "Yeah, it may not look pretty, but let's see some fool try to steal these bikes *now*," I chuckled to myself upon re-entering the traffic flow.

I spent the rest of the long drive westward reflecting back upon my personal journey—how much of it was spent on cruise control, blind to the things around me, in relentless pursuit of the next destination. I reckon the broken bike rack was a reminder that **I am now driven by purpose, not just accomplishments, and that redirections are gifts of guidance, not derailment.** I'm not going to give the impression it's easy; it's an ongoing practice, a process of learning how to listen to and follow your own intuition.

For me, that inner voice spoke freely as a child when I was immersed in nature. Climbing trees, wading in the creek, riding my horse, catching fireflies, planting crops, and exploring everything with my one-eyed Dalmatian, "Peppi." I felt an ongoing resonance and connectedness with my surroundings, an understanding and respect of the world and all of its living things around me. But somewhere along the way, that connectivity was lost.

Reconnecting wasn't easy for me. It took reaching a point of total burnout. In 2016, I walked away from an executive career, gave away most of my things, and lived for months in a remote area of Costa Rica. The burnout wasn't caused by a single event; it was a culmination over many years of allowing my power to be drained and doing things that didn't resonate. I felt out of sync because the things that were important to me seemed insignificant to those around me. There was so much pressure to follow the *right* path, I didn't realize *all* paths offered equal opportunity to invent, then reinvent, my own definition of "success."

Deep within the jungle, there was no phone service, no TV or radio, and no Wi-Fi, yet I was more connected there than I had been for decades. I reconnected with the everything-ness of nature, and realized that life is not about choosing sides: being logical *or* intuitive, rational *or* emotional, scientific *or* spiritual. It's about having the freedom to embrace *all* those qualities at the same time and tuning in to what's most important.

The lessons we need most—for both our personal balance and to power our modern organizations forward—have existed and thrived right under our feet since the beginning of time. **The living things around us have survived because they have evolved to their current state through millions of adaptations, just as we humans have done as well. Reconnecting to the natural world gave me the clarity to understand that reality is the intersection where quantum science meets purpose.** It is now clear that my gift is to help people and companies thrive by sharing a unique, timeless perspective on growth that stems from equal parts nature lover and business innovator.

Tuning In to Intuition

In the years since my jungle immersion, I've followed my intuition and listened to its deep wisdom when it told me when I was on the right path and when I was not. I credit that inner wisdom with leading me to this little town in the Rocky Mountains, where I can step outside into nature and easily do the things that energize me like hiking, riding horses, hot yoga, and biking. (Oh, and buying cool-smelling, natural deodorant recently made the high-vibe list, as well.) While my life was good in Nashville, there's a powerful force springing from deep inside me here by the mountains; something beyond the logical, three-dimensional world we can see and touch. It's a feeling of alignment with myself and the things around me, like I'm going twice the distance with half the effort.

For much of my life, I defined "success" from the outside in, which allowed me to gain financial resources and status, but not fulfillment. Now, I'm tuning in to that ancient, natural, never-ending, and ongoing source of power that's generated from the inside out, which creates a much greater form of personal success. It's no coincidence that world-famous innovators and billionaires Elon Musk, Oprah Winfrey, Richard Branson, and Steve Jobs all attribute their success to following their intuition. And with practice, anyone can hear their intuition and follow it, as well.

First, check your inputs. Energetically, science proves our thoughts and emotions drive our health and actions. Just like paying attention to the foods, beverages, and drugs you ingest, it's equally important to pay attention to the things you're tuning in to because they influence your thoughts and actions. Your current state is a direct reflection of the inputs you're allowing in to your life. What and who are you listening to? Is it your favorite social media app(s), the endless emails, or the 24/7 news drama? Is it the person who is the loudest or most outrageous, the most like you, or the most popular? Are the things and people around you raising your vibe or lowering it? To use an old programming phase, "garbage in, garbage out," meaning if we are constantly digesting negativity from people or things, we will begin to resonate with that vibe. On the other hand, if we can restrict access to the things that drain our energy and focus more on things that lift it, we'll be primed to hear our intuition more clearly.

In nature, owls are experts at tuning in to what's important. Their superpower is the ability to hyperfocus their vision and hearing. Their heads have adapted over the years to be shaped like satellite dishes that funnel information to their senses. Even their feathers are cupped to direct sound into their ears, and their heads swivel nearly all the way around to avoid blind spots. They are completely tuned in to the world around them.

Once you have assessed your inputs, the next step is to recognize and boost your vibe. To "tune in" to what your intuition is telling you, pay attention to the high-vibe things you're doing when you find yourself subconsciously humming, getting goosebumps, smiling, or having an extra bounce in your step. Those are the activities you are doing when you lose track of time, often referred to as being in a state of "flow."

In contrast, low-vibe activities can be felt in your body just above your navel. It's that pit in your stomach that tells you things are "off," like when someone you don't want to deal with calls and you see their name on the caller ID. You might freeze for a second or two before deciding whether to answer or not. That frozen feeling (of inertia) is an opportunity to recognize a low-vibe trigger.

Tuning into your intuition is easier to do when practiced habitually. Once you can recognize a high- or low-vibe state, you can use energy dynamics to balance and raise your vibe for a greater sense of well-being. Activities or thoughts that raise your vibration are called amplifiers, and the words used to describe them are expressed in energetic terms, such as being in sync, resonating, spun up with joy, lit up, transformed, accelerated, amplified, ignited, positive, tuned in, and connected.

Intuition makes you aware of amplifiers as directional indicators. Everyone tunes in differently, but some common activities that help are meditation (which can include various practices, such as prayer, yoga, art, or being immersed in nature), practicing gratitude, hiking, cycling, spending time with a pet, playing an instrument, or listening to your favorite music. I use little amplifiers, when possible, to gamify my workday and stay balanced by setting up small rewards like a quick walk around the block, eating a few bites of dark chocolate, or making a hot cup of tea.

Author Elizabeth Gilbert hosted a podcast where she talked about following your curiosity instead of your passion. I love that concept because it suggests that we honor our natural guidance system; the one that piques our desire to do, learn, or explore more of something. As great as passion can be, sometimes it can be disastrous. In my case, I was so passionate about escaping burnout that I gave away all my things, put my house on the market, and moved to the jungle. While it wound up working out, curiosity may have been a wiser choice!

For my logically minded readers, Lynn A. Robinson presents a useful technique in her book, *Put Your Intuition to Work*. She suggests framing anything you'd like your intuition's guidance on into a yes or no question. For example, "Should I ask for a raise? Does this person have my best interests in mind? Is this the right place to pack up and move?" Then flip a coin. Heads means "Yes" and tails means "No." The logic-based brain follows the coin. Your mind will be focused on observing the results, which allows your intuition a tiny window of time to sneak in and give you the answer before the coin lands. The split second it lands, notice your immediate thought. Is it "That's great!" or is it "Maybe I should make it the best two out of three?" The more you practice this simple technique, the more you can take note of how your body feels the vibe in that moment. Over time, you won't need the coin at all to feel your intuition; you'll be naturally attuned to it.

Plugging into the Supercharger

On October 31, 2022, a *Washington Post* headline read, "US Workers have gotten way less productive. No one is sure why." I beg to differ. Companies today expect employees to be highly motivated in the workplace, yet the pace of change and increasing work-life imbalance create uncertainty and frustration—both low-vibe emotions.

> If we want people to "be more productive," we need to provide them with the opportunity to recharge, which will happen more quickly when they do things that naturally energize them and align with their values. This is the prerequisite for bringing balance to the workplace.

The value of the 3-D AEM Cube to measure how people get into action using three *quantitative* axes (Exploration, Attachment, and Managing Complexity) cannot be understated. I have shared many examples of its transformational capabilities for teams of all levels and sizes in these pages. But there's another equally transformational way to supercharge people into action and that is by plugging into their Purpose, with a capital P. Purpose gets a higher output, faster; so much so that I call it the "fourth dimension" of success because it is *qualitative* and unseen, yet significantly enhances results.

There are many books on the topic of purpose in business. What I'm adding to that conversation is the recognition that Purpose is a natural amplifier or accelerant for Human Dynamics. Because energy itself is the great equalizer, everyone has the same capacity to become supercharged by plugging into Purpose. Like pouring gasoline on a flame, Purpose ignites engagement, retention, and productivity. According to the 2016 Deloitte Millennial Survey, 73% of employees who say they work at a "purpose-driven company" are engaged, compared to just 23% of those who don't see their company as purpose driven.

A word of caution: Do not mistake happiness for purpose. The chemical in the brain often associated with happiness is dopamine. It is a neurotransmitter that plays a role in reward-motivated behavior, pleasure, and reinforced learning. When you experience something pleasurable, such as eating your favorite ice cream, listening to music, or spending time with loved ones, your brain releases dopamine, which leads to feelings of happiness. However, happiness is a complex emotion that involves many different neurotransmitters and brain regions, including serotonin, oxytocin, endorphins, and the prefrontal cortex. These chemicals and brain regions work together to create our experiences of happiness and well-being.

It may seem counter-intuitive, but by focusing on becoming happy, we are constantly drawing our energy to the awareness that we are *not* happy. This actually lowers the chemicals in the brain associated with feeling upbeat, and causes people to have a lower, not higher, vibe. Studies in Tom Rath's book *Are You Fully Charged?* discuss this phenomenon. He says, "It's important to realize that purpose can drive happiness, but it doesn't work the other way around." As the book points out, purpose is an ongoing motivator that amplifies results for engagement, well-being, and productivity.

And a quick caveat for those who are unsure what their purpose is:

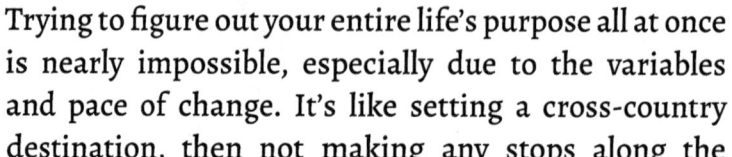

Trying to figure out your entire life's purpose all at once is nearly impossible, especially due to the variables and pace of change. It's like setting a cross-country destination, then not making any stops along the way to refuel and rejuvenate. Instead, we must break purpose up into waypoints: the short-term aspirations set for a period of time.

For example, moving to the mountains in Colorado is a waypoint in my journey. I'm not sure if I'll be here forever, but I do currently have a consistently higher vibe because the activities that energize me are easily accessible. Whereas, an aspiring singer-songwriter may view Nashville, aka "Music City," as a waypoint on their journey, surrounded by musicians, artists, and studios. The point is, everyone needs to pay attention to their own intuition to assess whether or not they are moving closer toward finding or living their purpose. This depends largely on how what you're doing—in your free time or in your career—makes you feel.

In the Industrial Era, the word *feel* was a 4-letter F-word, in the truest sense, because acknowledging that people had feelings was a surefire way to hinder efficiency and productivity. In the Human Era, where creativity and connection are paramount, how people *feel* is a critical driver of the modern workforce, especially with younger generations. Internal employees and external customers are constantly monitoring a company's practices, products, and values and they are widely communicating their assessments.

PURPOSE: THE FOURTH DIMENSION OF SUCCESS

Today's workers value purpose-driven roles more highly than traditional financial compensation, and once people are aligned with their purpose, they add value at a higher rate to the organization than those who are just looking for a paycheck. According to a report by Imperative, these purpose-oriented workers are more likely to be high performers and leaders in their organizations. They also report higher levels of well-being and fulfillment in their work, according to Imperative's 2018 State of the Purposeful Workforce. Companies that find ways to support personal purpose will reap the rewards through increased satisfaction, engagement, and ultimately retention. This translates directly into higher value and sustained growth.

A study by the *Harvard Business Review*, called "The Power of Purpose at Work," affirms that employees who feel their work is meaningful and aligned with their personal values are more than three times as likely to stay with their organizations. Another study conducted by BetterUp, "The State of Purpose at Work," shows that employees who identified with their company's purpose reported 20% higher job satisfaction and were 1.4 times more engaged than those who did not.

One simple way companies can energize employees is by encouraging each to set a personal goal for the year alongside their standard corporate metrics. Everyone has an activity that brings them a bit of purpose. It could be building a model plane, learning to play an instrument, taking an art class, hiking, training for a marathon…the list goes on and on. The goal can be tied to the organization in some way, like volunteering, or completely separate depending on each company's flexibility.

Executives have to stop thinking they can prescribe what will make their employees happy—a company picnic, a corporate outing, a tribal shout of the corporate mantra. While those can each be beneficial and fun activities, they aren't specifically relevant to each employee. The key is allowing the freedom for everyone to identify something that *they* are passionate about and also benefits them or the organization. Then give each person the flexibility and time to do it.

I realize this requires trust, but isn't that what we gave employees when they worked from home during the pandemic? Ask yourself: *What's the worst thing that can happen if an employee needs to leave early for guitar lessons once a week? Or what will happen if someone needs to take extended lunches to train for that marathon three times a week? Is it worth a few hours here and there to truly support someone's personal purpose?*

Yes. It is! According to studies recently published in the *Harvard Business Review*, over the last decade, purpose-driven brands have seen their valuation skyrocket by 175%. In one study of 28 companies over a 17-year period, purpose-driven companies grew by 1,681% in comparison with the S&P 500 average of 118% over the same span. Even more compellingly, 77% of consumers feel a stronger connection to purpose-driven companies over traditional companies, and 66% of consumers would switch from a product they typically buy to a new product from a purpose-driven company.[22]

As each person's connection to purpose increases, the overall business culture will shift. It has been reimagined as part of something bigger—something meaningful and relevant that shapeshifts it into the future of work.

With each new goal, each passion project over time, your company undergoes a *purposeful* transformation, which is good for people and good for business.

Conclusion
BE THE CHANGE

As I reflect on the pages of this book, I can't help but feel a sense of excitement and hope. The challenges we face in today's fast-paced business world can be overwhelming, but within each of us lies an incredible power waiting to be unleashed. This powerful current flows through every organization, every team, and every individual, ready to ignite our potential and propel us toward success.

I wrote this book with the heartfelt intention of sharing a transformative guide, one that connects the wisdom of the natural world with the practices of modern leadership and business. I hope you are inspired to embark on your own voyage of discovery, with the secrets of bioscience and Human Dynamics as your resource to unlocking the full potential of yourself and your workforce.

It's time to break free from the worn-out leadership methods of the past and the allure of fleeting trends. Instead, we can embrace the timeless principles found in nature's intricate systems—the complexity, physics, and ethology that govern everything in this world.

By understanding and leveraging these principles, we can adapt more effectively to the ever-accelerating pace of change and create thriving, innovative organizations.

Within these chapters lie insights and tools to navigate the challenges of imbalance, burnout, and the demands of a rapidly changing landscape. I hope you are inspired by the infinite power of energy—the life force that pulses through every individual, team, and organization. By tapping into this energy, we can unlock a new level of engagement, productivity, and innovation that ignites an extraordinary workplace culture where people and purpose align to amplify profits.

What's most exciting, *you have the power to be the change.* The knowledge and insights contained in these pages are not meant to be kept in theory but to be put into action. By observing and listening to the wisdom of the natural world, you can spark a positive movement within your organization and beyond. You can power the changes desperately needed to breathe new life into tired company cultures. I invite you to tap into your own power, to connect with those around you in a new and profound way, and to lead from within the current. The current, that ever-present electricity, hums within the heartbeats of your existing workforce and ignites every idea; it's the source of the energy and inspiration needed to get things done. It is at the core DNA of every living thing on the planet, from organisms to organizations. **The current starts small, hidden within our intuition, and spreads wide, manifesting into the emotions and actions that make us uniquely human. Like a seed, it grows into waypoints of purpose that can transform individuals, teams, organizations, and communities from drained and lifeless to connected, directed, and empowered to shine.**

Thank you for joining me on this journey. I have no doubt that you can also tap into nature's ultimate, timeless, and truly renewable resource—human energy. Let us ignite the potential within ourselves and those around us, transforming our organizations and making a lasting positive impact for generations to come.

ACKNOWLEDGEMENTS

Thanks first to my family for your ongoing love and support: Bobby and Edwanna Scott, Trent, Ethan, and Chris.

Next, I could not have completed this book without the "dream team" who came together to add great value to my writing. Brooke White, my writing partner and editor, consistently improved and clarified my ideas, adding her expertise as a business book writer and ultimately as a trusted friend. Martha Bullen, renowned publishing advisor, led our team through the process from manuscript to book launch and pulled together an A-list team to assist. Tracy Grigoriades, thank you for the overall project support, contract, and marketing guidance; and, David Aretha, your copy editing and suggestions are greatly valued.

Special thanks for concept conversations and thought leadership over the years to Neil Rodgers, Paradoxon Ltd., and Richard Lucy, former business colleague and partner. To content contributors Steve Keith, Richard Robertson, Sebastian Hamers, K'Leetha Gilbert Thomas, Leah Crabtree Gardner, and Gail Martino your insights are so appreciated. And, as always, I appreciate the concept reviews by Mike Modrak, JB Langston, and Jeff Stough.

SOURCES CITED

[1] "How many people have smartphones in 2023?" *Oberlo*. https://www.oberlo.com/statistics/how-many-people-have-smartphones

[2] Galov, Nick. "How Fast is Technology Growing—Can Moore's Law Still Explain the Progress?" *Web tribunal*, March 4, 2023. https://webtribunal.net/blog/how-fast-is-technology-growing/#gref

[3] Puiu, Tibi. "Your smartphone is millions of times more powerful than the Apollo 11 guidance computers." *ZME Science*, May 13, 2021. https://www.zmescience.com/feature-post/technology-articles/computer-science/smartphone-power-compared-to-apollo-432/

[4] McSpadden, Kevin. "You Now have a Shorter Attention Span than a Goldfish." *Time*, May 14, 2015.

[5] Rigoni, Brandon, Ph.D and Asplund, Jim. "Developing Employees' Strengths Boosts Sales, Profit, and Engagement." *Harvard Business Review*, September 1, 2016.

[6] "Gartner Says the Number of U.S. Employees Going Above and Beyond at Work at All-Time Low." *Gartner*, March 28, 2018. https://www.gartner.com/en/newsroom/press-releases/2018-03-28-gartner-says-the-number-of-us-employees-going-above-and-beyond-at-work-at-all-time-low

[7] Perl, Peter. "What is the Future of Truth?" *Pew*, February 4, 2019. https://www.pewtrusts.org/en/trust/archive/winter-2019/what-is-the-future-of-truth

[8] "The Problem with Burnout." *SHRM*. https://www.shrm.org/hr-today/news/hr-magazine/0817/pages/infographic-the-problem-with-burnout.aspx

[9] Mcleod, Saul, Ph.D. "Maslow's Hierarchy of Needs." *SimplyPsycyhology*, October 2, 2023. https://www.simplypsychology.org/maslow.html

[10] Vakil, Tushar. "What Makes Teams Successful? Google's Project Aristotle Came up with These Five Factors that Matter." NewAgeLeadershi, https://newageleadership.com/what-makes-teams-successful-googles-project-aristotle-came-up-with-these-five-factors-that-matter/#5_Factors_common_to_effective_teams_at_Google

[11] Scott, Tabitha, CEM, CDSM and Martha Amram, Ph.D. "Switch for Good Community Program Technical Report." *OSTI*, November 19, 2013. https://www.osti.gov/biblio/1123876

SOURCES CITED

[12] Bejan, Adrian. *The Physics of Life*. St. Martin's Press, 2016, Chapter 7.

[13] Holewinski, Britt. "Underground Networking: The Amazing Connections Beneath Your Feet." *National Forest Foundation*. https://www.nationalforests.org/blog/underground-mycorrhizal-network

[14] Yih, David. "Food, Poison, and Espionage: Mycorrhizal Networks in Action." *Arnoldia*, Vol. 75, Issue 2, November 15, 2012.

[15] Corpuz, Mark Gareth. "Cradle to Cradle: Principles, Examples, Pros, and Cons." *Profolus*, May 29, 2021. https://www.profolus.com/topics/cradle-to-cradle-principles-examples-pros-and-cons/

[16] "CCPII," Cradle-to-Cradle Products Innovation Institute, 2020. http://www.c2c-centre.com/sites/default/files/Shaw.pdf

[17] Robertson, Peter. *Always Change a Winning Team*. Marshall Cavendish Business, 2005.

[18] Robertson, Peter P. and Schoonman, Wouter. "How People Contribute to Growth-Curves." *SSRN*, July 8, 2013. https://papers.ssrn.com/sol3/papers.cfm?abstract_id=2291325

[19] "What is complex systems science?" Santa Fe Institute. www.santafe.edu. Archived from the original on April 14, 2022.

[20] "The AEM-Cube in practice." *human insight*. https://human-insight.com/strategic-solutions/aem-cube/

[21] Moore, Geoffrey. *Crossing the Chasm: Marketing and Selling High-Tech Products to Mainstream Customers*. Harper Business, 1991.

[22] Carucci, Ron and Ridge, Garry. "How Executive Teams Shape a Company's Purpose." *Harvard Business Review*, November 3, 2022.

Hi folks,

Just a quick personal note to thank you for purchasing and reading *Powering Change!* Ten percent of all profits will be used to benefit and support Indigenous Peoples and the natural world.

If you enjoyed this book, I invite you to:

- Get tips to implement the ideas from *Powering Change* for your own business at www.TabithaAScott.com.

- Share *Powering Change* with your colleagues and friends on social media, and leave a review online at your bookseller, like Amazon, to help others decide if it's right for them.

Your support means so much to me!

Warm regards,
Tabitha

ABOUT THE AUTHOR

TABITHA A. SCOTT is a business futurist, catalyst for positive change, global speaker, and bestselling author. With executive leadership in three $10+ billion global organizations, she has a proven track record in organizational transformation, innovation, and sustainable solutions through her powerful insights, engagements, and keynotes. Tabitha led efforts in creating the world's largest solar-powered community and was recognized for her technology innovations, including early AI deployment, by the White House.

Tabitha is the author of *Powering Change* and *Trust Your Animal Instincts*, which was awarded the prestigious Nautilus Literary Award for its positive impact on society. She often speaks about avoiding burnout, building positive cultures, modern forms of diversity (cognitive thinking and technologies), and how natural laws and systems inform smart business practices.

In addition to earning a BS in Finance, an MBA, and a Masters in Bank Management, Tabitha is a credentialed as a Certified Energy Manager and Certified Demand Side Manager through the Association of Energy Engineers, in Blockchain through MIT, and holds numerous certifications in human biofield holistic therapy. A Kentucky native, she currently lives near Boulder, Colorado.

Please visit www.TabithaAScott.com to learn more about her background and expertise, request an interview, or book her as a speaker.

Printed in the USA
CPSIA information can be obtained
at www.ICGtesting.com
LVHW051113290224
772930LV00017B/888

9 781735 494043